T0262275

Heat Exchangers: Methods and Design

Heat Exchangers: Methods and Design

Edited by **Edgar Miller**

New York

Published by NY Research Press,
23 West, 55th Street, Suite 816,
New York, NY 10019, USA
www.nyresearchpress.com

Heat Exchangers: Methods and Design
Edited by Edgar Miller

© 2015 NY Research Press

International Standard Book Number: 978-1-63238-286-3 (Hardback)

Printed in the United States of America.

Contents

Preface

This book offers valuable information on specific cases of heat exchangers. The selection was directed by seeking future prospects of applied research and industry, particularly aiming on the effective use and conversion energy in shifting environment. Beside the questions of thermodynamic basics, the book also presents various critical topics, such as operations, conceptions, design, fouling and cleaning of heat exchangers. It also addresses utilization and storage of thermal energy and geothermal energy, directly or by function of heat pumps. The contributions of this material are thematically grouped for the content of each topic, which is introduced by summarising the key objectives of the encompassed chapters. The book is not necessarily focused to be a fundamental source of knowledge in the area it covers, but rather serves as a mentor while practising expansive solutions of particular technical issues which are faced by engineers and technicians occupied in research and development in the subjects of heat transfer and heat exchangers.

Significant researches are present in this book. Intensive efforts have been employed by authors to make this book an outstanding discourse. This book contains the enlightening chapters which have been written on the basis of significant researches done by the experts.

Finally, I would also like to thank all the members involved in this book for being a team and meeting all the deadlines for the submission of their respective works. I would also like to thank my friends and family for being supportive in my efforts.

Editor

Part 1

Plate Heat Exchangers

The Characteristics of Brazed Plate Heat Exchangers with Different Chevron Angles

Muthuraman Subbiah
Higher College of Technology,
Oman

1. Introduction

Plate heat exchangers (PHEs) were introduced in the 1930s and were almost exclusively used as liquid/liquid heat exchangers in the food industries because of their ease of cleaning. Over the years, the development of the PHE has generally continued towards larger capacity, as well as higher working temperature and pressure. Recently, a gasket sealing was replaced by a brazed material, and each thermal plate was formed with a series of corrugations (herringbone or chevron). These greatly increased the pressure and the temperature capabilities.

The corrugated pattern on the thermal plate induces a highly turbulent fluid flow. The high turbulence in the PHE leads to an enhanced heat transfer, to a low fouling rate, and to a reduced heat transfer area. Therefore, PHEs can be used as alternatives to shell-and-tube heat exchangers. Due to ozone depletion, the refrigerant R22 is being replaced by R410A (a binary mixture of R32 and R125, mass fraction 50 %/50 %). R410A approximates an azeotropic behavior since it can be regarded as a pure substance because of the negligible temperature gliding. The heat transfer and the pressure drop characteristics in PHEs are related to the hydraulic diameter, the increased heat transfer area, the number of the flow channels, and the profile of the corrugation waviness, such as the inclination angle, the corrugation amplitude, and the corrugation wavelength. These geometric factors influence the separation, the boundary layer, and the vortex or swirl flow generation. However, earlier experimental and numerical works were restricted to a single-phase flow. Since the advent of a Brazed PHE (BPHE) in the 1990s, studies of the condensation and/or evaporation heat transfer have focused on their applications in refrigerating and air conditioning systems, but only a few studies have been done. Much work is needed to understand the features of the two-phase flow in the BPHEs with alternative refrigerants. Xiaoyang *et al.*, [1] experimented with the two-phase flow distribution in stacked PHEs at both vertical upward and downward flow orientations. They indicated that non-uniform distributions were found and that the flow distribution was strongly affected by the total inlet flow rate, the vapor quality, the flow channel orientation, and the geometry of the inlet port Holger [2].Theoretically predicted the performance of chevron-type PHEs under single-phase conditions and recommended the correlations for the friction factors and heat transfer coefficients as functions of the corrugation chevron angles. Lee *et al.*, [3] investigated the characteristics of the evaporation heat transfer and pressure drop in BPHEs with R404A and

R407C. Kedzierski [4] reported the effect of inclination on the performance of a BPHE using R22 in both the condenser and the evaporator. Several single-phase correlations for heat transfer coefficients and friction factors have been proposed, but few correlations for the two-phase flow have been proposed. Yan *et al.*, [5] suggested a correlation of condensation with a chevron angle of 30° for R134a. Yan *et al.*, reported that the mass flux, the vapor quality, and the condensation pressure affected the heat transfer coefficients and the pressure drops. Hieh and Lin [6] developed the correlations for evaporation with a chevron angle of 30° for R410A.

The main objective of this work was to experimentally investigate the heat transfer coefficients and the pressure drops during condensation of R410A inside BPHEs. Three BPHEs with different chevron angles of 45°, 35°, and 20° were used. The results were then compared to those of R22. The geometric effects of the plate on the heat transfer and the pressure drop were investigated by varying the mass flux, the quality, and the condensation temperature. From the results, the geometric effects, especially the chevron angle, must be considered to develop the correlations for the Nusselt number and the friction factor. Correlations for the Nusselt number and the friction factor with the geometric parameters are suggested in this study.

Experiments to measure the condensation heat transfer coefficient and the pressure drop in brazed plate heat exchangers (BPHEs) were performed with the refrigerants R410A and R22. Brazed plate heat exchangers with different chevron angles of 45°, 35°, and 20° were used. Varying the mass flux, the condensation temperature, and the vapor quality of the refrigerant, we measured the condensation heat transfer coefficient and the pressure drops. Both the heat transfer coefficient and the pressure drop increased proportionally with the mass flux and the vapor quality and inversely with the condensation temperature and the chevron angle.

Correlations of the Nusselt number and the friction factor with the geometric parameters are suggested for the tested BPHEs. In an effort to study and optimize the design of a plate heat exchanger comprising of corrugated walls with herringbone design, a CFD code is employed. Due to the difficulties induced by the geometry and flow complexity, an approach through a simplified model was followed as a first step. This simple model, comprised of only one corrugated plate and a flat plate, was constructed and simulated. The Reynolds numbers examined are 400, 900, 1000, 1150, 1250 and 1400. The SST turbulence model was preferred over other flow models for the simulation.

The case where hot water (60°C) is in contact with a constant-temperature wall (20°C) was also simulated and the heat transfer rate was calculated. The results for the simplified model, presented in terms of velocity, shear stress and heat transfer coefficients, strongly encourage the simulation of one channel of the typical plate heat exchanger, i.e. the one that comprises of two corrugated plates with herringbone design having their crests nearly in contact. Preliminary results of this latter work, currently in progress, comply with visual observations.

In recent years, compact heat exchangers with corrugated plates are being rapidly adopted by food and chemical process industries, replacing conventional shell-and-tube exchangers. Compact heat exchangers consist of plates embossed with some form of corrugated surface pattern, usually the chevron (herringbone) geometry[1].The plates are assembled being abutting, with their corrugations forming narrow passages. This type of equipment offers high thermal effectiveness and close temperature approach, while allowing ease of inspection and cleaning [1],[2]. In order to be able to evaluate its performance, methods to predict the heat transfer coefficient and pressure drop must be developed. In this direction, CFD is considered an efficient tool for momentum and heat transfer rate estimation in this type of heat exchangers.

The type of flow in such narrow passages, which is associated with the choice of the most appropriate flow model for CFD simulation, is still an open issue in the literature. Due to the relatively high pressure drop, compared to shell-and-tube heat exchangers for equivalent flow rates, the Reynolds numbers used in this type of equipment must be lower so as the resulting pressure drops would be generally acceptable[1]. Moreover, when this equipment is used as a reflux condenser, the limit imposed by the onset of flooding reduces the maximum Reynolds number to a value less than 2000[3]. Ciofalo et al.[4], in a comprehensive review article concerning modeling heat transfer in narrow flow passages, state that, for the Reynolds number range of 1,500-3,000, transitional flow is expected, a kind of flow among the most difficult to simulate by conventional turbulence models.

On the other hand, Shah & Wanniarachchi[1] declare that, for the Reynolds number range 100-1500, there is evidence that the flow is already turbulent, a statement that is also supported by Vlasogiannis et al.[5], whose experiments in a plate heat exchanger verify that the flow is turbulent for Re>650. Lioumbas et al.[6], who studied experimentally the flow in narrow passages during counter-current gas-liquid flow, suggest that the flow exhibits the basic features of turbulent flow even for the relatively low gas Reynolds numbers tested (500<Re<1200).Focke & Knibbe[7] performed flow visualization experiments in narrow passages with corrugated walls. They concluded that the flow patterns in such geometries are complex, due to the existence of secondary swirling motions along the furrows of their test section and suggest that the local flow structure controls the heat transfer process in such narrow passages.

The most common two-equation turbulence model, based on the equations for the turbulence energy k and its dissipation ε, is the k-ε model[8]. To calculate the boundary layer, either "wall functions" are used, overriding the calculation of k and ε in the wall adjacent nodes[8], or integration is performed to the surface, using a "low turbulent Reynolds (low-Re) k-ε" model[9]. Menter & Esch[9] state that, in standard k-ε the wall shear stress and heat flux are over predicted (especially for the lower range of the Reynolds number encountered in this kind of equipment) due to the over prediction of the turbulent length scale in the flow reattachment region, which is a characteristic phenomenon occurring on the corrugated surfaces in these geometries. Moreover, the standard k-ε, model requires a course grid near the wall, based on the value of $y+=11$ [9],[10], which is difficult to accomplish in confined geometries. The low-Re k-ε model, which uses "dumping functions" near the wall[8],[9], is not considered capable of predicting the flow parameters in the complex geometry of a corrugated narrow channel[4], requires finer mesh near the wall, is computationally expensive compared to the standard k-ε model and it is unstable in convergence.

An alternative to k-ε model, is the k-ω model, developed by Wilcox[11]. This model, which uses the turbulence frequency ω instead of the turbulence diffusivity ε, appears to be more robust, even for complex applications, and does not require very fine grid near the wall[8].

However, it seems to be sensitive to the free stream values of turbulence frequency ω outside the boundary layer. A combination of the two models, k-ε and k-ω, is the *SST* (Shear-Stress Transport) model, which, by employing specific "blending functions", activates the Wilcox model near the wall and the k-ε model for the rest of the flow[9] and thus it benefits from the advantages of both models. Some efforts have been made wards the effective simulation of a plate heat exchanger. Due to the modular nature of a compact heat exchanger, a common practice is to think of it as composed of a large number of unit *cells* (Representative Element Units, *RES*) and obtain results by using a single cell as the computational domain and imposing periodicity conditions across its boundaries[4],[12]. However, the validity of this assumption is considered another open issue in the literature [4].

2. Experimental facility

The experimental facility is capable of determining in plate heat transfer coefficients and measuring the pressure drops for the refrigerants. It consists of four main parts: a test section, a refrigerant loop, two water loops, and a data-acquisition system. A schematic of the test facility used in this study is shown in Figure-1, and detailed descriptions of the four main parts are mentioned below.

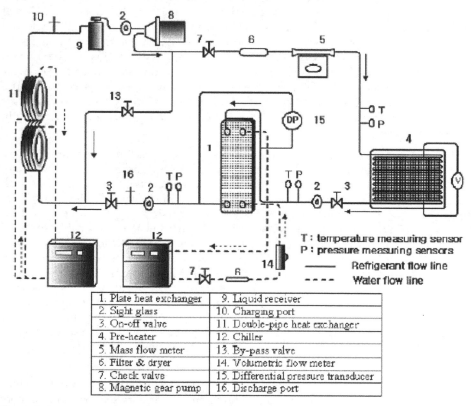

1. Plate heat exchanger	9. Liquid receiver
2. Sight glass	10. Charging port
3. On-off valve	11. Double-pipe heat exchanger
4. Pre-heater	12. Chiller
5. Mass flow meter	13. By-pass valve
6. Filter & dryer	14. Volumetric flow meter
7. Check valve	15. Differential pressure transducer
8. Magnetic gear pump	16. Discharge port

Fig. 1. Schematic diagram of the experimental system.

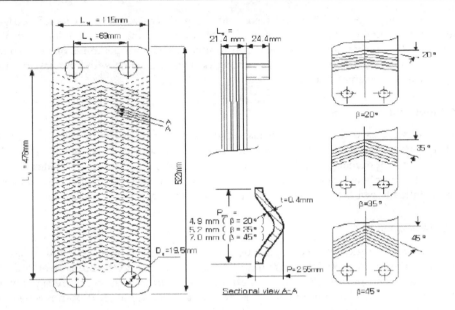

Fig. 2. Dimensions of the brazed plate heat exchangers.

2.1 Brazed plate heat exchangers

Three BPHEs with chevron angles of 45°, 35°, and 20° were used as the test sections. The angles of corrugation were measured from the horizontal axis. Each BPHE was composed of 4 thermal plates and 2 end plates, forming 5 flow channels. The dimensions of the BPHEs are shown in Figure-2. The refrigerant and cooling water were directed into the alternate passages between the plates through corner ports, creating counter flow conditions. The cooling water owed from the bottom to the top of every other channel on the basis of a central channel. On the other hand, the refrigerant owed from the top to the bottom in the rest of them.

2.2 Refrigerant loop

Refrigerant was supplied to the test section at specific conditions (i.e., temperature, flow rate, and quality) through the refrigerant loop. This loop contained a pre-heater, a double-pipe heat exchanger, a receiver, a magnetic gear pump, a differential pressure transducer, and a mass flow meter. Also included were thermocouples probes and pressure taps at the inlet/outlet of the test section. The refrigerant pump was driven by a DC motor which was controlled by a variable DC output motor controller.

The refrigerant flow rate was measured by using a mass flow meter installed between the magnetic gear pump and the pre-heater with an accuracy of 0.5 %. The pre-heater located before the test section was used to evaporate the refrigerant to a specified vapor quality at the inlet of the test section. The pressure drop of the refrigerant owing through the test section was measured with the differential pressure transducer, to an accuracy of 0.25 kPa. The refrigerant through the test section was subcooled at a double-pipe heat exchanger by the water cooled by the chiller and went into a liquid receiver. The subcooled refrigerant returned to the magnetic gear pump and circulated through the refrigerant loop repeatedly. Calibrated T-type thermocouples were used to measure the temperatures of the refrigerant at the inlet/outlet of the test section. The entire loop was insulated with fiberglass to prevent heat transfer to the environment.

2.3 Water loop

There are two closed water loops in this facility. One is for determining the condensation heat flux at the test section. The other is for making the subcooled refrigerant state at two double-pipe heat exchangers before it enters the magnetic gear pump. The water flow rates of the test section were measured by using a turbine flow meter, and T-type thermocouples were installed to evaluate the gain of the heat flux of the water of the test section.

2.4 Data acquisition

The data were recorded by a computer-controlled data-acquisition system with 40 channels scanned at the speed of 30 data per minute. The temperature and the pressure of both fluids were continuously recorded, and the thermodynamic properties of the refrigerant were obtained from a computer program. After steady-state conditions had been reached in the system, all measurements were taken for 10 minutes.

3. Data reduction and uncertainty analysis

The hydraulic diameter of the channel, D_h, is defined as

$$D_h = 4 \times \text{Channel flow area/Wetted perimeter} = 4bL_w/2L_w \emptyset \tag{1}$$

Where is $\emptyset$ =1.17. This value is given by the manufacturer.

The mean channel spacing, b, is defined as

$$b = p - t; \quad t = \text{Plate Thickness} \tag{2}$$

and the plate pitch p can be determined as, $\quad N_t$ = Total Number of plates

$$p = L_c/N_t - 1 \tag{3}$$

The procedures to calculate the condensation heat transfer coefficient of the refrigerant side are described below. At first, the refrigerant quality at the inlet of the test section (x_{in}) should be selected to evaluate the condensation heat at a given quality. Its value is calculated from the amount of heat given by a pre-heater, which is the summation of the sensible heat and the latent heat:

$$Q_{pre} = Q_{sens} + Q_{lat}$$

$$= m_r C_{p,r}(T_{r,sat} - T_{r,pre,in}) + m_r i_{fg} x_{in} \tag{4}$$

The refrigerant quality at the inlet of the test section can be written as

$$X_{in} = 1/i_{fg} [Q_{pre}/m_r - C_{p,r} (T_{r,sat} - T_{r,pre,in})] \tag{5}$$

The power gained by the pre-heater is calculated by measuring the voltage and the current with a power meter. The change in the refrigerant quality inside the test section was evaluated from the heat transferred in the test section and the refrigerant mass flow rate (6)

$$\Delta x = x_{in} - x_{out} = Q_w / m_r X i_{fg} \tag{6}$$

The condensing heat in the test section was calculated from an energy balance with water:

$$Q_w = m_w C_{p,w} (T_{w,out} - T_{w,in}) \tag{7}$$

The heat transfer coefficient of the refrigerant side (hr) was evaluated from the following equation:

$$1/h_r = (1/U) - (1/h_w) - R_{wall} \tag{8}$$

The overall heat transfer coefficient was determined using the log mean temperature difference

$$U = Q_w / A \times LMTD$$

$$LMTD = (T_{r,out} - T_{w,in}) - (T_{r,in} - T_{w,out}) / \ln\{(T_{r,out} - T_{w,in}) - (T_{r,in} - T_{w,out})\} \tag{9}$$

The heat transfer coefficient of the water side (h w) was obtained by using Eq. (10). Equation (10) was developed from the single-phase water to water pre-tests by K_{im} [7]. If the least-squares method and the multiple regression method are used, the heat transfer coefficient of the water side is correlated in terms of the Reynolds number, the Prandtl number, and the chevron angle:

$$h_w = 0.295(k_w / D_{Eq}) Re^{0.64} Pr^{0.32} (\pi/2 - \beta)^{0.09} \tag{10}$$

The thermal resistance of the wall is negligible compared to the effect of convection.

For the vertical downward flow, the total pressure drop in the test section is defined as

$$\Delta P_{total} = \Delta P_{fr} + \Delta P_a + \Delta P_s + \Delta P_p \tag{11}$$

And ΔP_{total} is measured by using a differential pressure transducer. The two-phase friction factor, f, is defined as

$$\Delta P_{fr} = f L_v N_{cp} G^2_{Eq} / D_h \rho_f \tag{12}$$

The port pressure loss in this experiment was less than 1 % of the total pressure loss. The static head loss can be written as and it has a negative value for vertical downward flow. The acceleration pressure drop for condensation is expressed as

$$\Delta P_p = 1.4 G^2_p / (2\rho_m) \tag{13}$$

An uncertainty analysis was done for all the measured data and the calculated quantities based on the methods described by Moffat [9]. The detailed results of the uncertainty analysis are shown in Table-1.

Parameters	Uncertainty
Temperature	$\pm 0.2\ ^{0}C$
Pressure	± 4.7 Pa
Pressure Drop	± 250 Pa
Water Flow Rate	$\pm 2\%$
Refrigerant mass flux	$\pm 0.5\%$
Heat flux of test section	$\pm 5.7\%$
Vapor Quality	± 0.03
Heat Transfer coefficients of water side	$\pm 10.1\%$
Heat transfer coefficients of refrigerant	$\pm 9.1\%$

Table 1. Estimated uncertainty.

Where

$$G_p = 4m_{Eq}/\pi D^2_p \tag{14}$$

And

$$(1/\rho_m) = (x/\rho_g) + [(1-x)/\rho_f]. \tag{15}$$

The equivalent mass flow rate, m_{eq}, is defined as

$$m_{eq} = m\ [1-x+x(\rho f/\ \rho g)^{0.5}] \tag{16}$$

The port pressure loss in this experiment was less than 1% of the total pressure loss. The static head loss can be written as

$$\Delta P_s = -\rho_m g L_v \tag{17}$$

And it has a negative value for vertical downward flow, The acceleration pressure drop for condensation is expressed as

$$\Delta P_a = - [(G^2_{eq}x/\rho_{fg})_{in} - (G^2_{eq}x/\ \rho_{fg})_{out}] \tag{18}$$

4. Results and discussions

The condensation heat transfer coefficients and the pressure drops of R410A and R22 were measured in three BPHEs with chevron angles of 20°, 35°, and 45° by varying the mass flux (13 - 34 kg/m²s), the vapor quality (0.9 - 0.15), and the condensing temperature (20°C and 30°C) under a given heat flux condition (4.7 -5.3 kW/m²). R22 was tested under identical experimental conditions for comparison with R410A.

4.1 Flow regime

Before the behaviors of heat transfer are considered, it is necessary to predict what flow regime exists at a given set of operating conditions. The detailed flow regime map for the PHE has not been proposed yet because of the difficulty of flow visualization. Vlasogiannis et al., [10] suggested the criterion of a two-phase flow regime for a PHE in terms of superficial liquid (jf) and vapor velocities (jg) by using water and air under adiabatic conditions. They only simulated a mixture of water and air as a two-phase fluid. According to their work, the flow patterns in a PHE are significantly different from those inside the vertical round tubes. They detected 3 types of flow patterns. The first was a gas continuous pattern with a liquid pocket at flow water flow rates (jf < 0.025 m/s) over wide range of air flow rates.

The second was the slug flow pattern, which was detected at sufficiently high air (jg > 2 m/s) and water flow rates (jf > 0.025 m/s). Thirdly, the liquid continuous pattern with a gas pocket or a gas bubble at the high water flow rates (jf >0.1 m/s) and low air flow rates (jg < 1 m/s).According to the flow regime map proposed by Vlasogiannis et al., the expected flow pattern in this experimental study is the gas continuous flow pattern with liquid pockets. However, their flow regime map has a significant limitation for use since many important features, such as the phase-change, the heating or cooling conditions, the densities or specific volumes of the working fluids, the geometries of the PHEs, etc., were not considered in detail. According to the flow regime map proposed by Crawford et al. [11], which was developed for vertical downward flow in a round tube, all experimental flow patterns are located in the intermittent flow regime, but this flow regime can not represent the correct flow regime in a BPHE due to the different geometries.

4.2 Condensation heat transfer

Figure-3 shows the effects of the refrigerant mass flux, the chevron angle, and the condensation temperature on the averaged heat transfer coefficient for R410A. The term "averaged heat transfer coefficient" means the average of the heat transfer coefficients calculated by varying the quality of the refrigerant from 0.15 to 0.9, and the coefficients were obtained from Eq. (19):

$$H_{averged} = \Sigma h_{local} x_{local} / \Sigma x_{local} \tag{19}$$

Where h_l is the local heat transfer coefficient at the local vapor quality. The experimental results indicate that the averaged heat transfer coefficients vary proportionally with the mass flux and inversely with the chevron angles and the condensation temperature. The small chevron angle forms narrow pitches to the flow direction, creating more abrupt changes in the velocity and the flow direction, thus increasing the effective contact length and time in a BHPE. The zigzag flow increases the heat transfer, and the turbulence created by the shape of the plate pattern is also important in addition to the turbulence created by the high flow rates. Increasing the mass flux at a given condensation temperature showed that the differences in the averaged heat transfer coefficients were significantly enlarged with decreasing chevron angle. This indicates that a PHE with the small chevron angle is more effective at a large mass flux (Gc > 25 kg/m²s) than at a small mass flux.

The averaged heat transfer coefficient of R410A decreases with increasing condensation temperature. The vapor velocity is a more influential factor than the liquid film thickness for the heat transfer. Vapor bubbles in the flow enhance the disturbance in the bubble wake as a turbulence promoter, and the turbulence induced by the vapor bubbles increases with the vapor velocity. Also, since the specific volume of the vapor increases with decreasing condensation temperature, the vapor velocity increases for a fixed mass flux and quality.

The vapor velocity at 20°C is faster than that at 30°C. The rates of the averaged heat transfer coefficients between condensation temperatures of 20°C and 30°C increased 5 % for a chevron angle of 45°, 9 % for 35°, and 16 % for 20°. These results show that different chevron angles lead partly to different flow pattern. Thus, we may conclude that the flow regime map should be modified by geometric considerations. The heat transfer coefficients in the high-quality region (fast velocity region) are larger than those in the low-quality region (slow velocity region). As mentioned above, this happens because the vapor velocity is the dominant effect on the heat transfer mechanism.

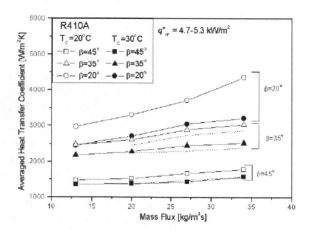

Fig. 3. Effect of mass flux on the averaged condensation heat transfer coefficient.

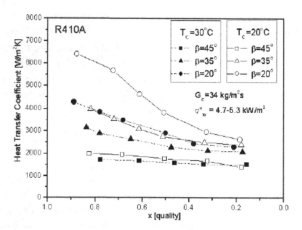

Fig. 4. Effect of quality on the condensation heat transfer coefficient.

Increasing the vapor quality at the same mass flux induces a faster bubble velocity, which increases the turbulence level and the convection heat transfer coefficient. The difference of heat transfer coefficients between the low-quality region and the high-quality region becomes larger with decreasing chevron angle. The PHE with a low chevron angle shows a better heat transfer performance in the high-quality region (i.e., the high vapor velocity region). Figure-4 also shows the variation of the heat transfer coefficients with the condensation temperatures. Like Figure-3, the heat transfer coefficients decreased with increasing condensation temperature. Also, the variations of the heat transfer coefficients with the condensation temperature are larger in the high-quality region. From the experimental results in Figures, 3 and 4, lowering the chevron angle and the condensation temperature gives the desired heat transfer effect.

4.3 Frictional pressure loss

The frictional pressure loss in a BPHE is obtained by subtracting the acceleration pressure loss, the static head loss, and the port pressure loss from the total pressure loss. Figure-5 shows the trend of the pressure drop along the mass flux, and Figure-6 shows the trend of the pressure drop along the quality at a mass flux of 34 kg/m²s and a heat flux of 4.7-5.3 kW/m². The frictional pressure drops in the BPHEs increase with increasing mass flux and quality and decreasing condensation temperature and chevron angle. This trend is similar to that of the condensation heat transfer. As mentioned above, since the vapor velocity is much faster than the liquid velocity during the two-phase flow in the tube, the vapor velocity is the dominant influence on the pressure drop, as well as the heat transfer. A high vapor velocity also tends to increase the turbulence of the flow. From Figures 3, 4, 5 and 6, we may concluded that since the trends of the the condensation heat transfer and the pressure loss in BPHEs are similar, those effects must be carefully considered in the design of a BPHE.

4.4 Comparison of R410A with R22

The ratios of R410A to R22 for the condensation heat transfer coefficients and pressure drops at a condensation temperature of 30°C are shown in the Figure-7. The ratios for the heat transfer coefficients are relatively constant in the range of 1 -1.1, regardless of the mass flux, while the ratios for the pressure drops decrease with increasing mass flux, except for the data at a chevron angle of 20° in the present experimental range. For a chevron angle of 20°, the heat transfer ratios of R410A to R22 are about 1.1, and the pressure drop ratios about 0.8, which is a 10 % higher heat transfer and a 20 % lower pressure drop.The smaller specific volume of the vapor of R410A relative to that of R22 makes the vapor velocity slower and yields a small pressure drop under the same conditions of the mass flux. While the two fluids have almost equal values of their latent heats, the liquid-phase thermal conductivity of R410A is larger than that of R22. The higher thermal conductivity for R410A helps to produce better heat transfer even if a reduction in the specific volume occurs. Also, a BPHE with a small chevron angle is known to have more effective performance from the ratios when replacing R22 with R410A.

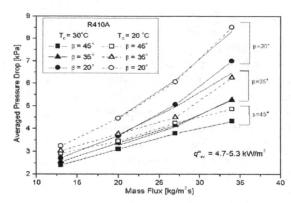

Fig. 5. Variation of the averaged condensation pressure drop with mass flux.

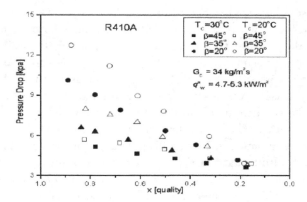

Fig. 6. Variation of the condensation pressure drop wih quality.

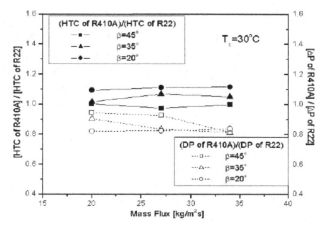

Fig. 7. Condensation heat transfer coefficient ratio and pressure drop ratio between R410A and R22.

4.5 Correlations of Nusselt number and friction factor for tested BPHEs

Based on the experimental data, the following correlations for Nu and f during condensation for the tested BPHEs are established: Where G_{e1}, G_{e2}, G_{e3}, and G_{e4} are non-dimensional geometric parameters that involve the corrugation pitch, the equivalent diameter, and the chevron angle. Re_{Eq} is the equivalent Reynolds number, and G_{Eq} the equivalent mass flux: where G_c is the channel mass flux. The suggested correlations for the Nusselt number and the friction factor can be applied in the range of Re_{Eq} from 300 to 4000. Figure-8(a) shows a comparison of the Nusselt number among the experimental data, the correlation proposed in this paper, and the correlation of Yan et al., [5]. The correlation of Yan et al., is

$$N_u = G_{e1}Re_{Eq}{}^{Ge1}Pr^{1/3} \tag{20}$$

$$Ge_1 = 11.22 \, (p_{co}/D_h)^{-2.83} \, (\Pi/2 - \beta)^{-4.5} \tag{21}$$

$$Ge_2 = 0.35 \, (p_{co} / D_h)^{0.23} \, (\Pi/2 - \beta)^{1.48} \tag{22}$$

$$f = Ge_3Re^{Ge4}{}_{Eq} \tag{23}$$

$$Ge_3 = 3521.1 \, (p_{co}/D_h)^{4.17} \, (\Pi/2 - \beta)^{-7.75} \tag{24}$$

$$Ge_4 = -1.024 \, (p_{co}/D_h)^{0.0925} \, (\Pi/2 - \beta)^{-1.3} \tag{25}$$

$$Re_{Eq} = G_{Eq}D_h / \mu_f \tag{26}$$

$$G_{Eq} = G_c[1-x+x(\rho_r/\rho_g)^{1/2}] \tag{27}$$

$$G_c = m / N_{ep}bL_w \tag{28}$$

and is obtained from one PHE with a chevron angle of 30° for R134a. Regardless of the BPHE types and refrigerants, most of the experimental data are within 20 % for the correlation proposed in this paper.

The correlation of Yan et al.(5), matched the data relatively well β for β: 20 and 35 within30 %, but over-predicted the data quite a bit for 45. This discrepancy results from the correlation of Yan et al., being developed for only a +30 PHE. Also,the correlation of Yan et al.

$$Nu = 4.118Re_{eq}{}^{0.4}Pr^{1/3} \tag{29}$$

for the Nusselt number only adopted the equivalent Reynolds number and Prandtl number without any geometric parameters. Because a BPHE has a strong geometric effect, the correlation with geometric parameters must be developed for general applications. The root-mean-square (r.m.s.) of the deviations is defined as

$$r.m.s. = \sqrt{1 / N_{data}\Sigma\left(Nu_{pred} - Nu_{exp} / Nu_{exp}\right)^2} \times 100(\%) \tag{28}$$

The r.m.s. deviation for the correlation of Yan et al., [Eq. (29)] is 50.2 % and for Eq. (20), it is only 10.9 %. Figure-8(b) shows a comparison of the friction factor between the experimental data and the proposed correlation. Similar to the correlation of the Nusselt number, the correlation of the friction factor includes the equivalent Reynolds number and the geometric parameters. Regardless of the BPHE types and refrigerants, most of the experimental data are within 15 % of the correlation proposed in this paper; the r.m.s. deviation for Eq. (23) is 10 %.

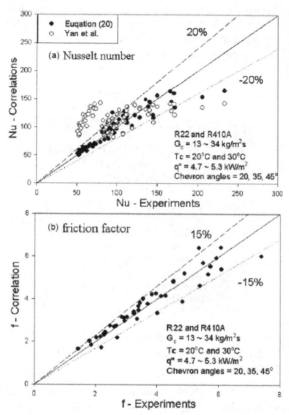

Fig. 8. Comparison of the correlations with the experimental data.

5. Study of a simplified geometry

In an effort to simulate the flow configuration, a **simple** channel was designed and constructed in order to conduct experiments and obtain formation on the flow pattern prevailing inside the furrows of the conduit. The flow configuration, apart from affecting the local momentum and heat transfer rates of a plate heat exchanger, suggests the appropriate flow model for the CFD simulation. A module of a plate heat exchanger is a single pass of the exchanger, consisting of only two plates. The simple channel examined is a single pass made of Plexiglas (**Figure 9**). It is formed by only **one** corrugated plate comprised of fourteen equal sized and uniformly spaced corrugations as well as a flat plate and it is used for pressure drop measurements and flow visualization. Details of the plate geometry are presented in **Table 2**. This model was chosen in an attempt to simplify the complexity of the original plate heat exchanger and to reduce the computational demands. The geometry studied in the CFD simulations (similar to the test section) is shown in **Figure 10**. The Reynolds numbers examined are 400, 900, 1000, 1150, 1250 and 1400, which are based on the distance between the plates at the entrance (d=10mm), the mean flow velocity and the properties of water at 60°C. In addition to isothermal flow, heat transfer simulations are carried out for the same Reynolds numbers, where hot water (60°C) is cooled in contact with a constant-temperature wall (20°C). The latter case is realized in condensers and evaporators. Additionally, it is assumed that heat is transferred only through the corrugated plate, while the rest of the walls are considered adiabatic.

Fig. 9. Simplified model and detail of the corrugated plate.

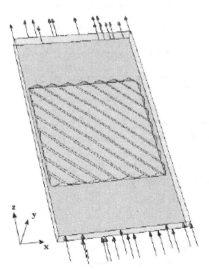

Fig. 10. CFD model.

A commercial CFD code, namely the *CFX ® 5.6* code developed by *AEA Technology*, was employed to explore its potential for computing detailed characteristics of this kind of flow. In general, the models used in CFD codes give reasonably good results for single-phase flow systems. The first step in obtaining a solution is the division of the physical domain into a solution mesh, in which the set of equations is discretised.

The grid size used is selected by performing a grid dependence study, since the accuracy of the solution greatly depends on the number and the size of the cells. The resulting mesh was also inspected for inappropriate generated cells (e.g. tetrahedral cells with sharp angles) and fixed, leading to a total number of 870,000 elements. The *SST* model was employed in the calculations for the reasons explained in the previous chapter. The mean velocity of the liquid phase was applied as boundary condition at the channel entrance (i.e. Dirichlet BC on the inlet velocity) and no slip conditions on the channel walls. A constant temperature boundary condition was applied only on the corrugated wall, whereas the rest of the walls are considered adiabatic. Calculations were performed on a *SGI O2* R10000 workstation with a 195MHz processor and 448Mb RAM. The *CFX ®5.6* code uses a finite volume method on a non-orthogonal body-fitted multi-block grid. In the present calculations, the *SIMPLEC* algorithm is used for pressure-velocity coupling and the *QUICK* scheme for discretisation of the momentum equations [31],[32].

Plate Length	0.200 m
Plate width	0.110 m
Maximum spacing between plates	0.010 m
Number of corrugations	14
Corrugation angle	45 °
Corrugation pitch	0.005 m
Corrugation width	0.014 m
Plate length before and after corrugations	0.050 m
Heat transfer area	$2.7 \times 10^{-2} \ m^2$

Table 2. Simple Channel's plate geometric characteristics.

The results of the present study suggest that fluid flow is mainly directed inside the furrows and follows them (*Figure 11a*). This type of flow behavior is also described by Focke & Knibbe[7], who made visual observations of the flow between two superposed corrugated plates (*Figure 11b*). They confirm that the fluid, after entering a furrow, mostly follows it until it reaches the side wall, where it is reflected and enters the anti-symmetrical furrow of the plate above, a behavior similar to the one predicted by the CFD simulation. It seems that, in both cases, most of the flow passes through the furrows, where enhanced heat transfer characteristics are expected.

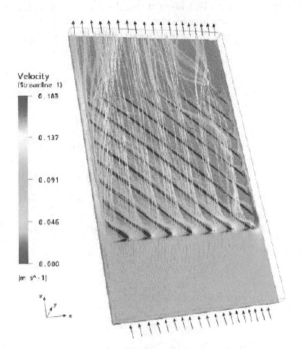

Fig. 11A. Typical flow pattern for the: a) simple channel, CFD results, Re=900.

Fig. 11B. Flow visualization by Focke & Knibbe[7], Re=125.

Figure 12 shows details of the flow inside a furrow for the simple model, where swirling flow is identified. This secondary flow is capable of bringing new fluid from the main stream close to the walls, augmenting heat transfer rates. Focke & Knibbe[18], who performed visualization experiments in similar geometries, also describe this kind of swirling flow. The values of the z-component of shear stress (**Figure13a**) increase with the Reynolds number –as expected–and the maximum value occurs at the crests of the corrugations. It may be argued that, during gas-liquid counter-current flow in such geometries, the shear stress distribution tends to prevent the liquid layer from falling over the crest of the corrugations and to keep it inside the furrows. The visual observations of Paras et al.[14] seem to confirm the above behavior. The heat flux through the wall of the corrugated plate was calculated by the CFD code. In addition, the local Nusselt number was calculated (by a user-Fortran subroutine) using the expression:

Fig. 12. Swirling flow inside a furrow; Re=900.

$$Nu_x = qd \: / \: (T_b - T_w) \: k \tag{31}$$

Where q' is the local wall heat flux, d the distance between the plates at the entrance, T_w the wall temperature, T_b the local fluid temperature and k the thermal conductivity of the fluid. In addition to the local Nusselt number, mean Nusselt numbers were calculated as follows:

* A *mean Nu* calculated by numerical integration of the local *Nu* over the *corrugated* area **only**, and

* An *overall* average *Nu* calculated using the total wall heat flux through the *whole* plate and the fluid temperatures at the channel entrance/exit.

The comparison of the values of the above Nusselt numbers shows that they do not differ more that 1%; therefore, the smooth part of the corrugated plate does not seem to influence the overall heat transfer. **Figure 13b** shows a typical local Nusselt number distribution over the corrugated wall for Re=900. All the Reynolds numbers studied exhibit similar distributions.

It is noticeable that local Nusselt numbers attain their maximum value at the top of the corrugations. This confirms the strong effect of the corrugations, not only on the flow distribution, but also on the heat transfer rate. To the best of author's knowledge, experimental values of heat transfer and pressure drop are very limited in the open literature for the corrugated plate geometry, since these data are proprietary. Therefore, the data of Vlasogiannis et al.[16] were used to validate the simulation results. These data concern heat transfer coefficients measurements of both single (Re<1200) and two-phase flow in a plate heat exchanger with corrugated walls and a corrugation inclination angle of 60o. Heavner et al.[14] proposed a theoretical approach, supported by experimental data, to predict heat transfer coefficients of chevron-type plate heat exchangers. **Figure 14** presents the experimental friction factors, obtained from the Plexiglas test section of **Figure 9**, as well as the CFD predictions for the simple geometry studied, as a function of the Reynolds number. It appears that the experimental values follow a power law of the form:

$$f = m\,Re^{-n} \tag{32}$$

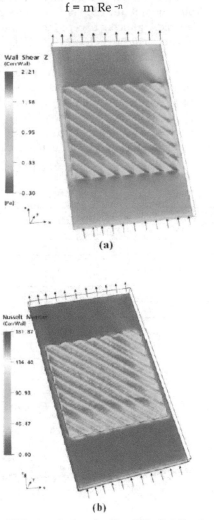

(a)

(b)

Fig. 13. Typical results of the CFD simulation for Re=900; distributions of: (a) z-shear stress component, (b) local Nusselt number.

Where m and n constants with values 0.27 and 0.14 respectively. Heavner et al.[14] proposed a similar empirical correlation based on their experimental results on a single pass of a plate heat exchanger with 45° corrugation angle, but with two corrugated plates. In spite of the differences in geometry, it appears that the present results are in good agreement with the experimental data of Heavner et al.[14] (0.687 and 0.141 for the variables m and n, respectively).

It must be noted that Focke et al.[15] , who also measured heat transfer coefficients in a corrugated plate heat exchanger having a partition of celluloid sheet between the two plates, reported that the overall heat transfer rate is the 65% of the corresponding value without the partition. **Figure 15** shows that the mean j-Colburn factor values calculated using the *overall* Nusselt number are practically equal to the 65% of the values measured by Vlasogiannis et al. This holds true for all Reynolds numbers except the smallest one (Re=400). In the latter case the Nusselt number is greatly overpredicted by the CFD code. This is not unexpected, since the *two-equation turbulence* model is not capable to predict correctly the heat transfer characteristics for such low Reynolds number.The CFD results reveal that the corrugations enhance the heat transfer coefficient, whereas the pressure losses due to the augmentation of friction factor f are increased **(Table 3)**, compared to a smooth-wall plate heat exchanger. Additionally, comparison of the normalized values of Nusselt number and the friction factor, with respect to the corresponding values for the smooth plate (fsm, $Nusm$), indicates that as the Reynolds number increases, heat transfer enhancement is slightly reduced, while the friction factor ratio, f/f , is increased. This is typical for plate heat exchangers with corrugations [16].

Re	Nu_{vlasog}	65% Nu_{vlasog}	Nu_{all}	Nu_{sm}	Nu_{ave}/Nu_{sm}	F/f_{sm}
400	13.2	8.6	20.5	-	-	-
900	38.0	24.7	27.3	9.4	2.9	12.4
1000	41.2	26.8	28.6	10.2	2.8	12.8
1150	44.2	28.7	28.8	11.0	2.7	13.5
1250	46.8	30.4	30.9	11.7	2.7	13.9
1400	49.5	32.2	32.0	12.5	2.6	14.5

Table 3. Experimental values, calculated Nusselt numbers and normalised values of N_u and f

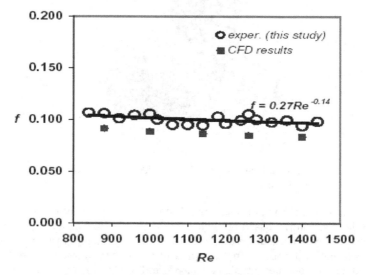

Fig. 14. Comparison of friction factor predictions (CFD) with experimental data.

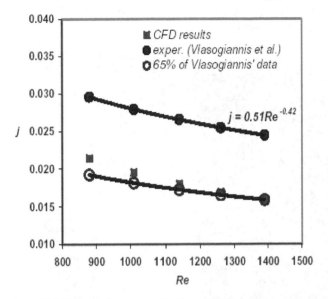

Fig. 15. Comparison of j-Colburn factor predictions (CFD) with experimental data.

6. Study of a heat exchanger channel

The results for the simplified geometry confirm the validity of the CFD code and strongly encourage the simulation of a module (pass) consisting of two corrugated plates of a compact heat exchanger (**Figure16a**). In order to quantitatively evaluate the results of this simulation, the experimental setup of Vlasogiannis et al.[16] was used as the design model (**Figure 16b**). Due to the increased computational demands, an AMD AthlonXP 1.7GHz workstation with 1GB RAM was used. The geometric characteristics of the new model are presented in **Table 4.**

Plate length	0.430 m
Plate width	0.100 m
Mean spacing between plates	0.024 m
Corrugation angles	60 °
Corrugation area length	0.352 m

Table 4. Geometric characteristics of the model with two corrugated plates.

Preliminary results of the present study, which is still in progress, are shown in **Figure 17**. It is obvious that the herringbone design promotes a symmetric flow pattern (**Figure 16b**). Focusing on the left half of the channel (**Figure 17a**), a close-up of the flow streamlines (**Figure17b**) reveals a *"peacock-tail"* pattern as the liquid flows inside the furrows and over the corrugations. The same flow pattern, which is characteristic for this type of geometry, has also been observed by Paras et al.[14] in similar cross-corrugated geometries (**Figure17c**), where "dry areas" of ellipsoidal shape are formed around the points where the corrugations come into contact. The effect of fluid properties (e.g. surface tension, viscosity) on the shape and the extent of these areas, which are considered undesirable, will be examined in the course of this study.

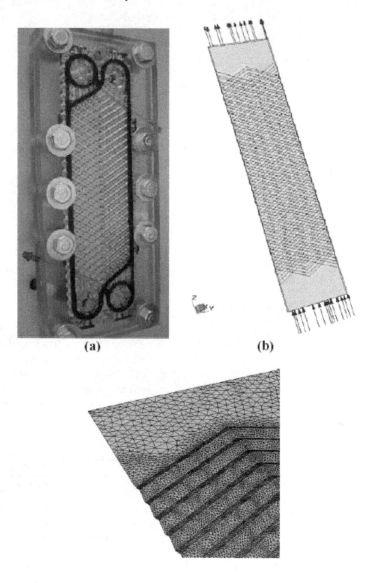

Fig. 16. (a) Module of a corrugated plate exchanger; (b) The CFD model and (c) Detail of the grid distribution over the corrugated wall.

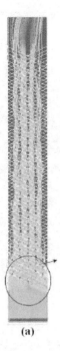

(a)

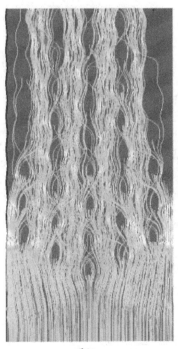

(b)

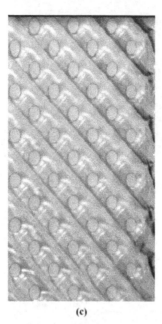

(c)

Fig. 17. (a) Streamlines in the left half of the channel; (b) Close up of the flow pattern; (c) Photo of the flow in the cross-corrugated geometry [14].

7. Conclusion

An experimental investigation has been conducted to measure the condensation heat transfer coefficient and the pressure drop of R410A and R22 in BPHEs with chevron angles of 20, 35, and 45 degrees. The experimental data were taken at two different condensation temperatures of 20°C and 30°C in the range of mass flux of 14-34 kg/m²s with a heat flux of 4.7 -5.3 kW/m².

- Both the heat transfer coefficient and the pressure drop increased proportionally with the mass flux and the vapor quality and inversely with the condensation temperature and the chevron angle. Those effects must be carefully considered in the design of a BPHE due their opposing effects.
- A comparison of the data for R410A and R22 showed that the heat transfer coefficient for R410A was about 0 - 10 % larger and the pressure drop about 2- 21 % lower than those for R22. Therefore, R410A is a suitable alternative refrigerant for R22.
- Correlations for the Nusselt number and the friction factor with the geometric parameters were suggested for the tested BPHEs within 20 % (r.m.s. deviation: 10.9 %) for Nu and 15 % (r.m.s. deviation: 10 %) for f.

Although compact heat exchangers with corrugated plates offer many advantages compared to conventional heat exchangers, their main drawback is the absence of a general design method. The variation of their basic geometric details (i.e. aspect ratio, shape and angle of the corrugations) produces various design configurations, but this variety, although it increases the ability of compact heat exchangers to adapt to different applications, renders it very difficult to generate an adequate 'database' covering all possible configurations. Thus, CFD simulation is promising in this respect, as it allows computation for various geometries, and study of the effect of various design configurations on heat transfer and flow characteristics.

In an effort to investigate the complex flow and heat transfer inside this equipment, this work starts by simulating and studying a simplified channel and, after gaining adequate experience, it continues by the CFD simulation of a module of a compact heat exchanger consisting of two corrugated plates. The data acquired from former simulation is consistent with the single corrugated plate results and verifies the importance of corrugations on both flow distribution and heat transfer rate. To compensate for the limited experimental data concerning the flow and heat transfer characteristics, the results are validated by comparing the overall Nusselt numbers calculated for this simple channel to those of a commercial heat exchanger and are found to be in reasonably good agreement. In addition, the results of the simulation of a complete heat exchanger agree with the visual observations in similar geometries.

Since the simulation is computationally intensive, it is necessary to employ a cluster of parallel workstations, in order to use finer grid and more appropriate CFD flow models. The results of this study, apart from enhancing our physical understanding of the flow inside compact heat exchangers, can also contribute to the formulation of design equations that could be appended to commercial process simulators. Additional experimental work is needed to validate and support CFD results, and towards this direction there is work in progress on visualization and measurements of pressure drop, local velocity profiles and heat transfer coefficients in this type of equipment.

8. Appendix

Nomenclature

A	heat transfer area of plate [m^2]
b	mean channel spacing [m]
C_p	constant pressure specific heat [J/kg K]
D	diameter [m]
f	friction factor
G	mass flux [kg/m^2s]
Ge	non-dimensional geometric parameter
g	gravitational acceleration [m/s^2]
h	heat transfer coefficient [W/m^2K]
i	enthalpy [J/kg]
j	superficial velocity [m/s]
L_c	distance between the end plates [m]
L_h	distance between the ports [m]
L_v	vertical length of the fluid path [m]
L_w	horizontal length of the plates [m]
LMTD	log mean temperature difference [°C]
m	mass flow rate [kg/s]
N_{cp}	number of channels for the refrigerant

N_{data} total number of data
N_t total number of plates
Nu   Nusselt number
Nu_{exp} Nusselt number obtained from experiment
Nu_{pred} Nusselt number obtained from correlation
p   plate pitch [m]
p_{co} corrugation pitch [m]
Pr   Prandtl number [v]
Q   heat transfer rate [W]
q   heat flux [W/m²]
Re   Reynolds number
T   temperature [°C]
t   plate thickness [m]
U   overall ht coefficient [W/m² K]
x   quality

Subscripts

a   acceleration
c   channel
Eq   equivalent
f   liquid
fg   difference the liquid phase and the vapor phase
fr   friction
g   vapor
in   inlet
lat   latent
m   mean
out   outlet
p   port
pre   pre-heater
r   refrigerant
s   static
sat   saturated
sens   sensible
w   water

9. References

[1] X. Rong, M. Kawaji and J.G. Burgers, Two-phase header flow distribution in a stacked plate heat exchanger, *Proceedings ASME/JSME FED-Gas Liquid Flows* 225 (1995), pp. 115–122.

[2] H. Martin, 1996, A theoretical approach to predict the performance of chevron-type plate heat exchangers, Chemical Engineering and Processing: Process Intensification, Volume 35, Issue 4, Pages 301-310.

[3] G. J. Lee, J. Lee C. D. Jeon and O. K. Kwon. 1999. In: Plate Heat Exchanger with chevron angles ,Proceedings of the 1999 Summer Meeting of the SAREK, edited by C. S. Yim (SAREK, Nov.). p. 144.

[4] M. A. Kedzierski. 1997. Heat Exchanger Multiphase flow, Heat Transfer Engineering. Volume 5, issue 3 page 18: 25.

[5] Y. Y. Yan, H. C. Lio and T. F. Lin. 1999. Different Chevron angles in plate heat exchanger, of Heat and Mass Transfer. Volume 11, issue 4 pages 42: 93

[6] Y. Y. Hsieh and T. F. Lin. 2002.plate heat exchanger design theory, International journal of Heat and Mass Transfer. Volume 21, issue 9 pages 1033-45.

[7] Y. S. Kim. 1999. Plate heat exchanger design, M.S. Thesis. Yonsei University.

[8] S. Kakac and H. Liu. 1998. Heat Exchangers Selection, Rating and Thermal Design. CRC Press, Boca Raton. Volume 8, issue 9 pages 323-329

[9] R. J. Mo. 1982. Model of plate heat exchanger, ASME Journal of fluid engineering, Volume 11, issue 9 pages 173-179

[10] P. Vlasogiannis, G. Karajiannis. 2002. Compact heat exchangers, International journal Multiphase Flow.21, issue 9 pages 728: 757.

[11] T. J. Crawford, C. B. Weinberger and J. Weisman. 1985. heat exchangers International journal Multiphase Flow.21, issue 9 pages 291: 297.

[12] Shah, R.K., Wanniarachchi, A.S. (1991), Plate heat exchanger design theory, In: Buchlin, J.-M. (Ed.),Industrial Heat Exchangers, von Karman Institute Lecture Series 1991-04.

[13] Kays, W.M. & London, A.L. (1998), Compact heat exchangers, 3rd Ed. Krieger Publ. Co., Florida.

[14] Paras, S.V., Drosos, E.I.P., Karabelas, A.J, Chopard, F. (2001), "Counter-Current Gas/Liquid Flow Through Channels with Corrugated Walls–Visual Observations of Liquid Distribution and Flooding", World Conference on Experimental Heat Transfer, Fluid Mechanics & Thermodynamics, Thessaloniki, September 24-28.

[15] Ciofalo, M. Collins, M.W., Stasiek, J.A. (1998), Flow and heat transfer predictions in flow passages of air preheaters: assessment of alternative modeling approaches, In: Computer simulations in compact heat exchangers, Eds. B. Sunden, M.Faghri, Computational Mechanics Publ. U.K.

[16] Vlasogiannis, P., Karagiannis, G., Argyropoulos, P., Bontozoglou, V. (2002), "Air–water two-phase flow and heat transfer in a plate heat exchanger", Int. J. Multiphase Flow, 28, 5, pp. 757-772.

[17] Lioumbas, I.S., Mouza, A.A., Paras, S.V. (2002), "Local velocities inside the gas phase in counter current two-phase flow in a narrow vertical channel", Chemical Engineering Research & Design, 80, 6, pp. 667-673.

[18] Focke, W.W., Knibbe, P.G. (1986), "Flow visualization in parallel-plate ducts with corrugated walls", J. Fluid Mech., 165, 73-77.

[19] Davidson, L. (2001), An Introduction to Turbulence Models, Department of Thermo and Fluid Dynamics, Chalmers University of Technology, Götemberg, Sweden.

[20] Menter, F., Esch, T. (2001), "Elements of Industrial Heat Transfer Predictions", 16th Brazilian Congress of Mechanical Engineering (COBEM), 26-30 Nov. 2001, Uberlandia, Brazil.

[21] AEA Technology (2003), CFX Release 5.6 User Guide, CFX International, Harwell, Didcot, UK.

[22] Wilcox,D(1988), "Reassessment of the scale-determining equation", AIAA Journal, 26,11.

[23] Mehrabian, M.A., Poulter, R. (2000), "Hydrodynamics and thermal characteristics of corrugated channels: computational approach", Applied Mathematical Modeling, 24, pp. 343-364.

Part 2

Energy Storage Heat Pumps Geothermal Energy

2

Ground-Source Heat Pumps and Energy Saving

Mohamad Kharseh

Willy´s CleanTech AB, PARK 124 Karlstad,
Sweden

1. Introduction

The global warming itself and its consequences cause considerable problems. It results in extreme climate events such as droughts, floods, or hurricanes, which are expected to become more frequent. This puts extra strain on people and has great impact on public health and life quality especially in poor countries.

Internationally, there is a political understanding that global warming (or climate change) is the main challenge of the world for decades to come. Thus, all states must work together in order to overcome climatic change consequences.

Although, studies suggest that there is indeed relationship between solar variability and global warming (Lean and Rind, 2001), two causes of the warming have been suggested:

1. related to the accumulation of greenhouse gases in the Earth's atmosphere;
2. related to heat emissions (Nordell, 2003, Nordell and Gervet, 2009).

This implies that current warming is anthropogenic and caused by human activities, i.e. global use of non-renewable energy. So far, the total global energy consumption has already exceeded $15 \cdot 10^{10}$ MWh/year and it is projected to have an annual growth rate about 1.4 % until 2020 (EIA, 2010).

Much of the energy used worldwide is mainly supplied by fossil fuels (~85 % of the global energy demand while renewable energy sources supply only about 6 %) (Moomaw et al., 2011, Jabder et al., 2011, Jaber et al., 2011). Owing to global dependence on oil fuels has resulted in a daily oil consumption of 87.7 million barrels (Mbbl), Fig. 1 (IEA, 2010, EIA, 2007). Consequently, about $3 \cdot 10^{10}$ ton of carbon dioxide emissions are annually emitted into the atmosphere. In other word, for each consumed kWh about 205 kg of carbon dioxide is being emitted into the atmosphere.

Environmental reasons urge us to find more efficient ways in converting and utilizing the energy resources. From the environment point of view, there is now almost universal scientific acceptance that profligate energy use is causing rapid and dangerous changes in the global climate. There is mounting evidence that the mean global temperature has increased over the period 1880 to 1985 by 0.5 to 0.7 °C (Hansen and Lebedeff, 1987). While surface air temperature (SAT) compilations shows that SAT has increased 1.2 °C last century. If a current climatic change trend continues, climate models predict that the average global temperature are likely to have risen by 4 to 6 ºC by the end of 21st century (Gaterell, 2005). As climate change progresses, all the other environmental problems are becoming worse and harder to solve. Therefore, a sustainable future requires worldwide efforts to prepare for new energy sources and a more efficient use of energy.

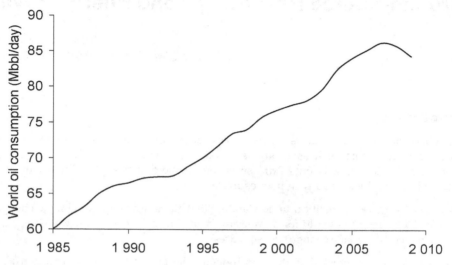

Fig. 1. World oil consumption.

Thanks to the awareness of the impact of global warming and its relationship with human activities, there has been a growing interest in reducing fossil energy consumptions. Specifically, more efficient use of energy and increased use of renewable energy seem to be our main weapon against the ongoing global warming.

Heating and cooling in the industrial, commercial, and domestic sectors accounts for about 40-50 % of the world's total delivered energy consumption (IEA, 2007, Seyboth et al., 2008). Although, buildings regulations aim to reduce the thermal loads of buildings, as the economic growth improves standards of living, the energy demand for heating and cooling is projected to increase. For example, in non-OECD nations, as developing nations mature, the amount of energy used in buildings sector is rapidly increasing. Consequently, the implementation of more efficient heating/cooling systems is of clear potential to save energy and environment. However, the use of renewable energy systems for heating and cooling applications has received relatively little attention compared with other applications such as renewable electricity or biofuels for transportation. Yet, renewable energy sources supply only around 2-3% of annual global heating and cooling (EIA, 2010). It is worth mentioning that a century or more ago renewable energy accounted for almost 100%. In other word, all current researches aim to approach what was the case in the past.

Nowadays, and due to its high thermal performance, the ground source heat pump (GSHP) has increasingly replaced conventional heating and cooling systems around the world. Such system extracts energy from a relatively cold source to be injected into the conditioned space in winter or alternatively, extracts energy from conditioned spaces to be injected into a relatively warm sink in summer.

Current work emphasizes the importance of using ground source heat pumps in reaching towards the renewable energy goals of climate change mitigation, and reduced environmental impacts.

2. Principle of GSHP systems

The ground source heat pump (GSHP) system are also known as ground coupled heat pump (GCHP), borehole systems or borehole thermal energy storage (BTES), and shallow geothermal system. Due to its high thermal performance, the ground source heat pump (GSHP) have increasingly replaced conventional heating and cooling systems around the world (IEA, 2007, Hepbasli, 2005, De Swardt and Meyer, 2001). Essentially GSHP systems refer to a combination of a heat pump and a system for exchanging heat from the ground. The GSHPs move heat from the ground to heat homes in the winter or alternatively, move heat from the homes to the ground to cool them in the summer. This heat transfer process is achieved by circulating a heat carrier (water or a water–antifreeze mixture) between a ground heat exchanger (GHE) and heat pump. The GHE is a pipe (usually of plastic) buried vertically or horizontally under the ground surface, Fig. 2 (Sanner et al., 2003). At the beginning of 2010 the totally installed GSHP capacity in the world was 50,583 MW producing 121,696 GWh/year with capacity factor and annual grow rate of 0.27 and 12.3%, respectively (Lund et al., 2010).

Heating mode: In this case, the GHE and the heat pump evaporator are connected together and the heat pump moves the heat from the ground into the conditioned space. The liquid of relatively low temperature is pumped through the GHE, collecting heat from the surrounding ground, and into the heat pump. Since the temperature of extracted liquid, which is around mean annual air temperature, is not suitable to be used directly for heating purpose, heat pump elevates the temperature to a suitable level (30-45 °C) before it is submitted to a distribution system.

Cooling mode: In this case, the GHE and the heat pump condenser are connected together and heat pump moves the heat from the conditioned space into the ground. The liquid of relatively high temperature is pumped through the GHE, dispersing heat into the surrounding ground, and into the heat pump.

As known, heat transfers from a warmer object to a colder one. Heat, as stated by the second law of thermodynamics, cannot spontaneously flow from a colder location to a hotter area unless work is done. The heat pump is simply a device for absorbing heat from one place and transporting it to another of relatively lower temperature. So, such device can be used to maintain a space temperature at desired level by removing unwanted heat (e.g. a fridge or air conditioning unit) or to transport heat to where it is wanted (space or water heating). In space conditioning application, heat pump system is composed of an indoor unite and an outdoor unite and the task of the heat pump is to transfer heat from one unite to the other. In order to keep inside temperature at comfort level in the winter, for example, the heat pump absorbs heat from outdoor and expels it into building. In the summer the reversed process occurs, i.e. the heat pump moves heat from indoor and expels it to outside.

The temperature difference between the indoor unite and outdoor unite is referred to as temperature lift. This temperature plays a major role in determining the coefficient of performance of heat pump (COP= delivered energy/driving energy). A smaller temperature

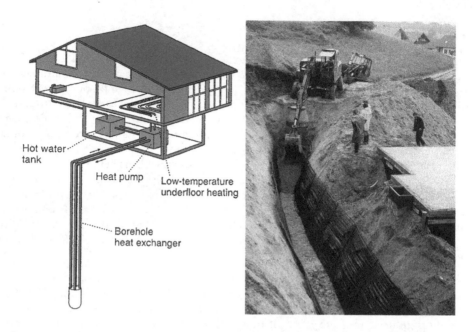

Fig. 2. Typical application of ground source heat pump system (Sanner et al., 2003).

lift results in a better COP. More specifically, extracting heat from a warmer medium during the heating season and injecting heat into a colder medium during cooling season leads to a better COP and, consequently, less energy use.

Fig.3 shows a schematic illustration of the components of assumed system as well as the thermodynamic cycle on diagrams temperature-entropy and pressure-enthalpy. Many techniques have been recently proposed in order to improve the cycle performance, more details are given by Wang, 2000, Chap.9 (Wang, 2000). In the current work, a heat exchanger has been added between the suction line and liquid line.

Like a heat engine but operating in reverse, the thermodynamics of the cycle can be analyzed on diagrams. In general COP is defined as the ratio between the delivered capacity and compressor capacity (Wang, 2000):

$$COP_c = \frac{Q_c}{Wcp} \tag{1}$$

$$COP_h = \frac{Q_h}{Wcp} \tag{2}$$

Where Q_h, Q_c, and W_{cp} represent the heating, cooling, and compressor capacity, respectively.

As shown in the Fig. 3, the heat exchanging operations in the evaporator and the condenser occurs at constant pressure processes (isobar). The compression process in the compressor befall at isentropic process theoretically, while the expansion operation in the expansion valve occurs at adiabatic process. With these in mind, as per the thermodynamics rules, the terms of Eq.1 and Eq.2 might be calculated as follows:

$$Q_h = m \cdot (h_3 - h_4) \tag{3}$$

$$Q_c = m \cdot (h_7 - h_6) \tag{4}$$

$$W_{CP} = m \cdot (h_2 - h_1) \tag{5}$$

Where, h and m represent enthalpy and refrigerant mass flow rate, respectively (see Fig. 3).

In order to accomplish the calculations, the following assumptions were made:

- Refrigerant R22
- Pressure drop at inlet and outlet of the compressor was assumed $P_8-P_1=10$
- and $P_2-P_3=23$ KPa respectively, see Fig.3.
- The pressure drop through the pipe is negligible.
- The isentropic efficiency of the compressor is 80%.
- There is no sub-cooling in the condenser or useless superheat in suction line.
- Thermal efficiency of the heat exchanger, which expresses how efficient the heat exchanger utilizes the temperature difference, is 90%
- Heat loss factor of the compressor, i.e. ratio between heat loss of the compressor to the surroundings and the energy consumption of the compressor, is 15%.
- Regarding to the internal unite of the heat pump, in wintertime the condensation temperature was assumed 38 ᴼC. In summertime, the evaporation temperature was assumed 8 ᴼC.
- Heating/Cooling capacity assumed to be constant, thus a change in temperature will affect the flow rate of refrigerant through the cycle.

The calculation results are illustrated in Fig.4. Apparently, the COP of heating machine increases as the evaporation temperature rises. Likewise, the performance of cooling machine increases as the condensation temperature decreases.

The ground temperature below a certain depth is constant over the year. This depth depends on the thermal properties of the ground, but it is in range of 10-15 m, see section 3 below. Thus, the ground is warmer than the air during wintertime and colder than the air during the summertime. Therefore, use the ground instead of the air as heat source or as a heat sink for the heat pump results in smaller lift temperature and, consequently, better thermal performance. In addition to improve the COP, the relatively stable ground temperature means that GSHP systems, unlike ASHP, operate close to optimal design temperature thereby operating at a relatively constant capacity. It is good to mention here that in outdoor unite fan, in ASHP case, consumes more energy than that of the water pump in the GSHP case (De Swardt and Meyer, 2001). Therefore, the comparison would be even more favorable for the GSHP, if the fan energy consumption is considered in the COP calculation.

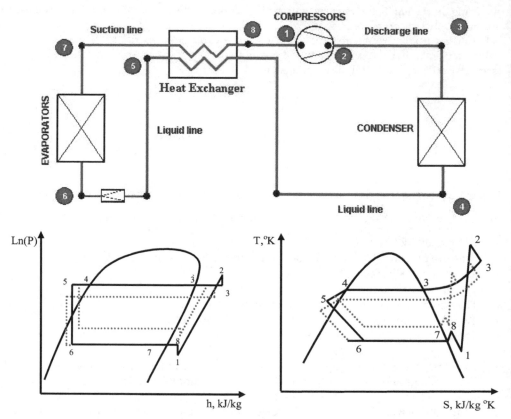

Fig. 3. Illustration of heat pump and the thermodynamic cycle on the LnP-h and T-S diagram.

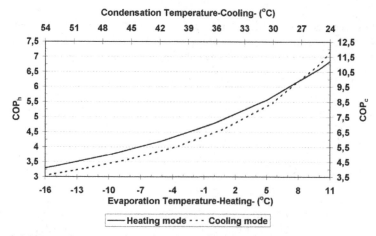

Fig. 4. Actual COP as a function of condensation/evaporation temperature.

3. Ground temperature

The ambient air temperature over the year or the day can be treated as a sinusoidal function around its average value Ta . This fluctuation might be expressed by:

$$T(t) = T_a + A_a \cdot \cos(2\pi \cdot \frac{t}{t_o}) \tag{6}$$

Where $T(t)$ is air temperature at given time t; T_a is average air temperature for the period t_o, A_a is the air temperature amplitude (half of the difference between the maximum and minimum temperatures for the period), t_o is the time for one complete cycle (day or year).

Apparently, air temperature fluctuation generates variations in the ground temperature. In order to find out a mathematical expression of ground temperature, the equation to be solved is the one-dimensional heat conduction equation. The mathematical formulation of this problem is given as:

$$\frac{\partial^2 T(z,t)}{\partial z^2} = \frac{1}{\alpha} \cdot \frac{\partial T(z,t)}{\partial t} \tag{7}$$

Where α is the thermal diffusivity (m²/s), z depth below the surface (m), t is the time. Note that for oscillating temperature at the boundary, we do not need an initial condition The solution of Eq.12 can be found by Laplace transformation method (Carslaw and Jaeger, 1959):

$$T(t,z) = T_a + A_a \cdot e^{-\frac{z}{d_o}} \cdot \cos(2\pi \cdot \frac{t}{t_o} - \frac{z}{d_o}) \tag{8}$$

Where d_o is the penetration depth (m), which is defined as the depth at which the temperature amplitude inside the material falls to 1/e (about 37%) of the air temperature at the surface:

$$d_o = \sqrt{\frac{\alpha \cdot t_o}{\pi}} \tag{9}$$

Fig.5 shows the underground temperature as function of the depth at different seasons of the year. As shown, below a certain depth, which depends on the thermal properties of the ground, the seasonal temperature fluctuations at ground surface disappears and ground temperatures is essentially constant throughout the year. In other word, for depth below a few meters ground is warmer than air during the winter and colder than the air during the summer.

Eq. 8 shows that ground temperature amplitude decreases exponentially with distance from the surface at a rate dictated by the periodic time, mathematically we can write:

$$A_g = A_a \cdot e^{\frac{-z}{d_o}} \tag{10}$$

Where A_g is ground temperature amplitude (°C).

In addition, Eq. 8 shows that there is a time lag between the ground and air oscillating temperature. In other words, the maximum or minimum ground temperature occurs later than the corresponding values at the surface. From the cosine term in Eq. 8 one can conclude that the time lag increasing linearly with depth. The shifting time, φ , between surface and the ground at a given depth z is:

$$\varphi = t_2 - t_1 = \frac{z}{2} \cdot \sqrt{\frac{C \cdot t_0}{\pi \cdot \lambda}} \qquad (11)$$

Indeed, change in temperature of ambient air results in change in the undisturbed ground temperature. Measurements of borehole temperature depth profile (BTDP) evidently show that there are temperature deviations from the linear steady-state ground temperature in the upper sections of boreholes (Goto, 2010, Harris and Chapman, 1997, Lachenbruch and Marshall, 1986, Guillou-Frottier et al., 1998). Mathematical models have been used to simulate the change in ground temperature due to GW. Kharseh derived a new equation that gives the ground temperature increase in areas where the surface warming is known (Kharseh, 2011). The suggested solution is more user-friendly than other solutions. The derived equation was used to determine the average change of ground temperature over a certain depth and therefore the heat retained by a column of earth during the warming period. This average change of ground temperature is of great importance in the borehole system.

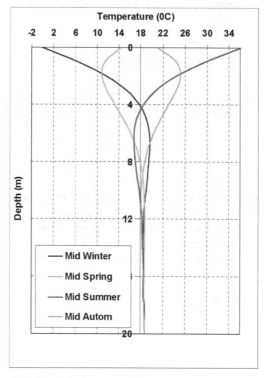

Fig. 5. Temperature profile through the ground.

4. Ground source heat pump systems and energy saving

4.1 Case study – the Kharseh chicken farm

The Kharseh chicken farm in Hama, Syria, was selected as a study case to show the contribution of ground source heat pumps in saving energy consumption of heating and cooling system. Even though the annual mean temperature in Syria is 15-18 °C, heating of such farm consumes considerable amounts of energy. The reason is that the air temperature is close to freezing during three winter months and that chickens require a relatively high temperature, 21-35 °C, depending on chickens' age as seen in Table. 1.

Age of chicken	Temperature at 0.10-0.15 m level
(weeks)	(°C)
1	35
2	32
3	29
4	27
5	24
6 (fully grown)	21

Table 1. Appropriate indoor temperature in chicken farms.

The chicken hangar is placed parallel to the main wind direction has a floor area of 500 m² (50 m x 10 m) in E–W direction. The total window area is 24 m².

4.2 Heating/cooling demand

The mean heating load composed of heat losses through the external walls and ventilation, while cooling load composed of heat gained through external walls, ventilation, solar radiation, and heat released by chickens. In current work the degree-hour method was used to estimate the thermal demand of the hangar (Durmayaz et al., 2000) using following assumptions:

- External wall's area, of thermal resistance 0.45 K.m²/W, is 336 m²
- Floor and ceiling area, of thermal resistance 5 and 0.45 K.m²/W, respectively, is 500 m²
- Windows's area, of thermal resistance 0.2 K.m²/W, is 24 m²
- Ventilation rate 20 m³/m²,h (ventilated area of chicken farm varies with chicken age)
- Heat release from chickens: 50 W/m² (varies with age)
- The capacity of the hangar is 5 cycles/year of 55 days the period of each cycle life. This mean that the hangar will be occupied 75% out of the entire year.
- During their first day, the chickens occupy about 85 m² of the building. This area is increased 14 m² per day until they occupy the entire area of the hangar after about one month. This mean the average occupied are during one cycle is 77% out of whole hangar's area.

- Heating season is 6 months, while cooling season is 4 months.
- 10 h of cooling and 24 hours of heating are required per a day during summer and winter, respectively.

Using these assumptions, the total heat loss coefficient of the hangar, L (W/K), can be calculated as follow:

$$L = \frac{(\rho C_p)_{air} \cdot I \cdot V}{3600} + \Sigma U \cdot A \tag{12}$$

Finally, annual heating demand, Q_h (MWh), is

$$Q_h = \frac{L \cdot DHh - 50 \cdot 500 \cdot 24 \cdot 30 \cdot 6}{10^6} \tag{13}$$

While the annual cooling demand, Q_c (MWh), is

$$Q_c = \frac{L \cdot DHc + 50 \cdot 500 \cdot 10 \cdot 30 \cdot 4}{10^6} \tag{14}$$

Where DHh and DHc is the total number of degree-hours of heating and cooling, respectively, which can be calculated as follow:

$$DHh = \sum_{j=1}^{N} (T_i - T_o)_j \qquad when\ is \quad T_o \leq T_b \tag{15}$$

While for cooling (DHc)

$$DHc = \sum_{j=1}^{K} (T_o - T_i)_j \qquad when\ is \quad T_o \geq T_b \tag{16}$$

Where T_b is the base temperature and T_i represents the indoor design temperature, T_o is the hourly ambient air temperature measured at a meteorology station, N is the number of hours providing the condition of $T_o \leq T_b$ in a heating season while K is the number of hours providing the condition of $T_o \geq T_b$ in a cooling season. In current work, and due to considering the big internal load, base temperature was assumed to be equal to T_i. Since the indoor temperature varies with the time during chickens cycle, the indoor temperature was assumed to be constant during one cycle and equals the average temperature i.e. T_i=28 °C. Fig.6 shows that the estimated total annual heating demand is 230 MWh while the corresponding cooling demand is 33 MWh.

In order to determine the maximum required heating and cooling capacity, the required heating/cooling power as the chickens grow during the hottest and coldest period of the year were calculate. As shown in Fig. 7, during heating season, due to lowering the appropriate indoor temperature with age and due to increase the occupied area, the heating power increases with time until it peaks in the middle of the chickens' life cycle. This peak demand does not occur during the cooling season. The calculations showed that the maximum required heating and cooling capacity are 113 kW and 119 kW, respectively.

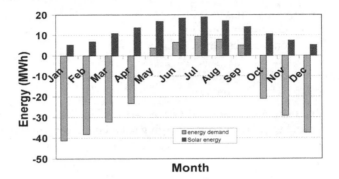

Fig. 6. Monthly heating/cooling demand and solar yield.

It should be noted that in Kharseh, 2009 the German DIN was used for the same aim. Therefore there is a small different in estimated thermal demand of the hangar.

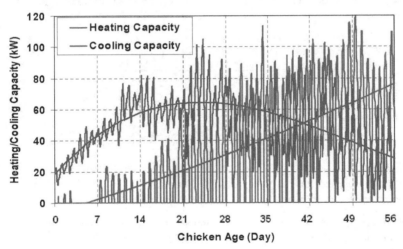

Fig. 7. Heating/cooling power as function of chicken age for one complete cycle during the hottest and coldest period.

4.3 System design and simulated operation

The EED (Earth Energy Design) model(EED, 2008) was used in pre-designing required borehole system to meet to estimated heating/cooling load at given conditions.

4.3.1 Borehole system

Specific data of the borehole system are given below:

- Number of boreholes: 10
- Borehole Diameter: 0.11 m
- Borehole Depth: 120 m

- Volumetric heat capacity: 2.16 MJ/m^3.K
- Ground thermal conductivity: 3.5 W/m.K
- Drilling Configuration: open rectangle 175 (3 x 4)
- Borehole Spacing: 6 m.
- Borehole installation: Polyethylene U-pipe
- Fluid flow rate: 0.5 10^{-3} m^3/s, borehole.

To keep the borehole temperature at steady state between the years extracted and injected heat from/to the ground were balanced by charging solar heat during the summer.

4.4 Solar collector

Since the annual heating demand of the hangar is much greater than annual cooling demand, which mean the energy extracted from the ground will be more than that injected into the ground, recharging the borehole filed by external energy resource is need. The amount of available solar energy in Syria means great potential for combined solar and GSHP systems. The estimated required solar collector area without considering heat yield from ground was:

$$A = \frac{Q_h \cdot (1 - \frac{1}{COP_h}) - Q_c \cdot (1 + \frac{1}{COPc})}{\eta \cdot \sigma} \tag{17}$$

Where

Q$_h$	Heating demand (MWh)
COP$_h$	Coefficient of performance for heating (in this case =5)
Qc	Cooling demand (MWh)
COP$_c$	Coefficient of performance for free cooling (in this case =50)
σ	Yearly sun yield (in this case σ=1.973 MWh/m2)
η	Solar collector efficiency (in this case η=0.86).

In this case, the required solar collector area was 85 m2. The solar heat is directly used when needed while the rest of the heat is stored to be used later (Fig.8).

4.5 Operation

During the wintertime Fig.8, water is pumped from the borehole through the solar collector to increase its temperature. The temperature increase, which is only 0.8 °C during the winter, is considerably greater during the summer. The heat pump cools the water before it is again pumped through the borehole, where it will be warmed up. The extracted heat is emitted into the hangar. Fig.9 shows that the lowest extracted water temperature from borehole is 11.5 °C. During summertime Fig.8, the ground temperature is cold enough for free cooling, so the water is pumped directly to the heat exchanger. Due to the heat exchange with indoor air, the water temperature will increase. After the heat exchanger, water passes though the solar collector and back to the borehole. Then, its temperature will decrease before pumped back to the hangar. Fig.9 shows that the highest extracted water temperature from borehole is 26.5 °C

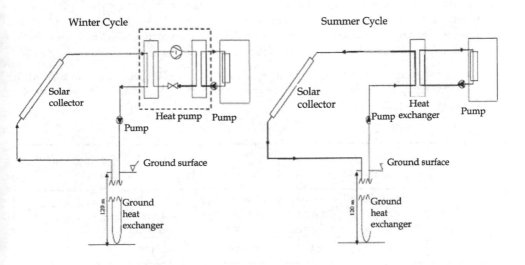

Fig. 8. Schematic of the solar coupled to ground source heat pump system.

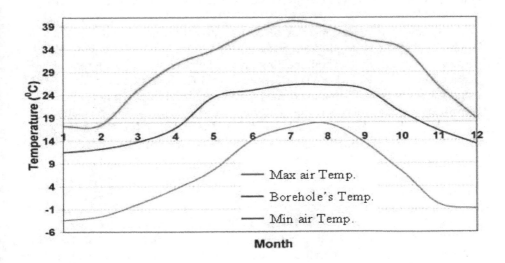

Fig. 9. Relevant temperatures for performed calculations.

5. Results and discussion

Present study was performed to determine the potential of GSHP, with solar collectors, for heating and cooling purposes in the Middle East. The Kharseh chicken farm in Syria of area 500 m² was chosen as a case study. The heating and cooling demands of the hangar were then used to estimate annual heating and cooling demands of the ideal chicken farm in Syria of area 200 m². The calculations showed the following results:

- A typical average size chicken house in Syria requires 92 MWh of heating and 13 MWh of cooling. Required heating and cooling powers are 45.2 kW and 47.7 kW, respectively, as shown in table 2.

Typical farm size Floor area 200 m²			Totally for 13000 farms Floor area 2.6 Mm²		
Meat production ton/year	Heating Energy MWh/year	Cooling Energy MWh/year	Total energy for heating GWh/y	Total energy for cooling GWh/y	Total Energy GWh/year
13	92	13	1196	170	1366

Table 2. Heating and cooling demand for chicken farms in Syria.

- 480 m of borehole with diameter 0.11 m and 34 m² of a solar collector were required to supply the heating and cooling of the typical chicken farms in Syria. In this case, the maximum fluid temperature delivered from the boreholes is 26.5 °C in the summer while the minimum mean fluid temperature was 11.5 °C during the winter.

- Table 3 shows the operation costs of coal furnace heating system combined with ASHP for cooling issue, diesel furnace heating system combined with ASHP for cooling issue, ASHP for both heating and cooling issue, and suggested GSHP heating/cooling system. Using Fig. 4, we found the COP_h and COP_c for the ASHP are 4 and 4.3, respectively, while the corresponding values for GSHP are 6.2 and 10. The conversion efficiency of conventional heater was assumed 85%. The calculations show that by using the GSHP, the annual operation costs can be reduced 38%, 69.2%, and 79.7 % compared to ASHP, coal heater combined with ASHP, and diesel heater combined with ASHP, respectively.

- Table 4 shows comparison between the required prime energy, i.e. tons of coal, of three different systems assuming the average annual efficiency of the power plant 32%. As shown, using the GSHP, the amount of fuel required is reduced 38% compared to ASHP or 57.2% compared to coal heater combined with ASHP. In other words, by use GSHP in all chicken farms in Syria, the annual coal consumption can be reduced $107.6 \cdot 10^3$ ton compared to traditional existing system (coal heater combined with ASHP). Accordingly, the carbon dioxide emission can be reduced by the same percentages.

Energy demand GWh/year	Energy Cost (MSP)			
	GSHP 3.5 SP/kWh	ASHP 3.5 SP/kWh	Coal Heater 8,141 kWh/kg, 13 SP/kg	Diesel Heater 10,1 kWh/l 25 SP/L
Heating	1196	675 (COP=6.2)	1047 (COP=4)	2247 (η=0.85)	3479(η=0.85)
Cooling	170	60 (COP=10)	138 (COP=4.3)	138 (COP=4.3)	138 (COP=4.3)
Total	1366	735	1185	2385	3617
Energy Cost SP/kWh		0.54	0.87	1.75	2.65

Table 3. Comparison between different heating/cooling systems for a typical chicken farm.

System	Required prime Energy GWh/y	Required Coal (10³ ton)
GSHP	210	80.6
ASHP	338	130
Coal Heater with ASHP	1446	188.2

Table 4. comparison between the required prime energy.

- The estimated installation cost of a borehole system for a typical chicken farm is $15000. With current energy price in Syria the payback-time of GSHP is about 5.3, or 3 years compared to coal heater combined with ASHP, or diesel heater combined with ASHP, respectively.

6. Conclusions

The global energy oil production is unstable and will peak within a few years. Therefore, the energy prices are expected to rise and new energy systems are needed. In addition to this energy crisis the fossil fuels seems to be the main reason for climate change. There is a global political understanding that we need to replace fossil fuels by renewable energy systems in order to develop a stable and sustainable energy supply.

About half of the global energy consumption is used for space heating and space cooling systems. Ground source heat pump systems are considered as an energy system that can make huge contributions to reduce energy consumption and thereby save the environment.

7. Nomenclature

Q_c	cooling demand, (MWh)
Q_h	heating demand, (MWh)
W_{cp}	compressor capacity, (kW)
COP_c	coefficient of performance of cooling mode, (dimensionless)
COP_h	coefficient of performance of heating mode, (dimensionless)
h	enthalpy, (kJ/kg.K)
P	pressure, (Pa)
m	refrigerant mass flow rate, (kg/s)
$T(t)$	air temperature at given time t, (K)
T_a	average air temperature, (K)
A_a	the air temperature amplitude, (K)
A_g	ground temperature amplitude, (K).
t_o	the time for one complete cycle (day or year) of air temperature variation
z	depth below the surface, (m)
α	ground thermal diffusivity, (m²/s)
d_o	penetration depth, (m)
φ	shifting time between the air and the ground temperatures variation,(s)
DH	degree-hour, (h.K)
CDH	cooling degree-hour, (h.K)
HDH	heating degree-hour, (h.K)
L	total heat loss coefficient of building, (W/K)
T_b	base temperature, (K)
T_o	outdoor temperature, (K)
Min T	minimum fluid temperature extracted from the borehole, (K)
Max T	maximum fluid temperature extracted from the borehole, (K)
σ	Yearly sun yield, (kWh/m²)
η	Solar collector efficiency

8. References

Carslaw, H. S. & Jaeger, J. C. 1959. *Conduction of heat in solids*, Oxford: Clarendon.

De Swardt, C. A. & Meyer, J. P. 2001. A performance comparison between an air-source and a ground-source reversible heat pump. *International Journal of Energy Research,* 25, 899-910.

Durmayaz, A., Kadioglu, M. & Sen, Z. 2000. An application of the degree-hours method to estimate the residential heating energy requirement and fuel consumption in Istanbul. *Energy,* 25, 1245-1256.

EED. 2008. *Earth Energy Designer* [Online]. Available: www.buildingphysics.com/earth1.htm [Accessed].

EIA. 2007. *Energy Information Administration* [Online]. Available:
www.eia.doe.gov [Accessed].

EIA 2010. International Energy Outlook 2010. Washington.

Gaterell, M. R. 2005. The impact of climate change uncertainties on the performance of energy efficiency measures applied to dwellings. *Energy and Buildings,* 37, 982-995.

Goto, S. 2010. Reconstruction of the 500-year ground surface temperature history of northern Awaji Island, southwest Japan, using a layered thermal property model. *Physics of the earth and planetary interiors,* 183, 435-446.

Guillou-Frottier, L., Mareschal, J.-C. & Musset, J. 1998. Ground surface temperature history in central Canada inferred from 10 selected borehole temperature profiles. *Journal of geophysical research,* 103, 7385-7397.

Hansen, J. & Lebedeff, S. 1987. Global Trends of Measured Surface Air-Temperature. *Journal of geophysical research: Atmospheres,* 92, 13345-13372.

Harris, R. N. & Chapman, D. S. 1997. Borehole temperatures and a baseline for 20th-century global warming estimates. *Science,* 275, 1618-1621.

Hepbasli, A. 2005. Thermodynamic analysis of a ground-source heat pump system for district heating. *International Journal Of Energy Research,* 29, 671-687.

IEA 2007. Renewables For Heating And Cooling. Paris, France: International Energy Agency.

IEA 2010. International Energy Agency. Oil and Gas Markets. Paris, France.

Jabder, S. A. A., Amin, A. Z., Clini, C., Dixon, R., Eckhart, M., El-Ashry, M., Fakir, S., Gupta, D. & Haddouche, A. 2011. Renewables 2011 Global Status Report. Paris, France: Renewable Energy Policy Netwrok for the 21st Century.

Jaber, S. A. A., Amin, A. Z., Clini, C., Dixon, R., Eckhart, M., El-Ashry, M., Fakir, S., Gupta, D. & Haddouche, A. 2011. Renewables 2011 Global Status Report. Paris, France: Renewable Energy Policy Netwrok for the 21st Century.

Kharseh, M. 2011. Ground Response to Global Warming. *Journal of geophysical research,* Submited.

Lachenbruch, A. H. & Marshall, B. V. 1986. Changing Climate: Geothermal Evidence from Permafrost in the Alaskan Arctic. *Science,* 234, 689-696.

Lean, J. & Rind, D. 2001. Earth's response to a variable sun. *Science,* 292, 234-236.

Lund, J. W., Freeston, D. H. & Boyd, T. L. Year. Direct Utilization of Geothermal Energy 2010 Worldwide Review. *In:* World Geothermal Congress 2010, 25-29 April 2010 2010 Bali, Indonesia. 1-23.

Moomaw, W., Yamba, F., Kamimoto, M., Maurice, L., Nyboer, J., Urama, K. & Weir, T. 2011. Introduction. In IPCC Special Report on Renewable Energy Sources and Climate Change Mitigation. Cambridge.

Nordell, B. 2003. Thermal pollution causes global warming. *Global and planetary change,* 38, 305-312.

Nordell, B. & Gervet, B. 2009. Global energy accumulation and net heat emission. *International Journal of Global Warming,* 1, 373-391.

Sanner, B., Constantine Karytsasb, Dimitrios Mendrinosb & Rybachc, L. 2003. Current status of ground source heat pumps and underground thermal energy storage in Europe. *Geothermics,* 32, 579-588.

Seyboth, K., Beurskens, L., Langniss, O. & Sims, R. E. H. 2008. Recognising the potential for renewable energy heating and cooling. *Energy Policy,* 36, 2460-2463.

Wang, S. K. 2000. *Handbook of air conditioning and refrigeration,* New York, McGraw Hill.

3

PCM-Air Heat Exchangers: Slab Geometry

Pablo Dolado, Ana Lázaro, José María Marín and Belén Zalba
University of Zaragoza / I3A - GITSE
Spain

1. Introduction

Energy efficiency and the search for new energy sources and uses are becoming main objectives for the scientific community as well as for society in general. This search is due to various environmental issues and shortages of conventional and non-sustainable energy resources, for example fossil fuels, that are essential to industrial development and to daily life. Free-cooling in buildings, bioclimatic architecture applications, demand and production coupling in renewable energy sources, as solar energy, are examples of thermal energy storage contributions to achieve these objectives. The application of Phase Change Materials (hereafter PCM) in Thermal Energy Storage (hereafter TES) is an expanding field due to the variety of materials being developed. There are four critical considerations for the technical viability of these applications: 1) The features of both the PCM and the encapsulation material must be stable during the system lifetime; 2) A reliable numerical model of the system to simulate different operational conditions; 3) The thermophysical properties of the PCM; 4) The cost of the system.

Specifically, the solid-liquid phase change phenomenon of the PCM is being widely studied in the field of TES, both experimentally and numerically, because this technology is of great interest among different fields: from applications in electronics, textile, transport... to applications in aerospace or thermo-solar power plants. The incorporation of these materials on the market, as stated before, is conditioned partly by its price. To cope with this situation, manufacturers often sell PCM as non-pure substances or mixtures which, on the one hand, lower their costs but, on the other hand, condition its thermophysical properties so that they are not as well established as in pure substances. Generally, this determining factor leads to a nonlinearity of the temperature dependence of the thermophysical properties of the PCM. This issue is another aspect to consider when simulating the thermal behaviour of these substances. Therefore, it is essential a good determination of these properties as they are input values to the theoretical models that simulate the thermal performance of devices based on these materials, some of which may strongly condition the results of the simulations.

When working at ambient temperatures, there are different situations where TES with PCM can be applied. Zalba et al., 2003, presented a comprehensive review on latent heat TES and its applications. The authors remarked that low values of λ_{PCM} can lead to real problems in the systems since there could be insufficient capacity to dispose of the stored energy quickly enough. Later, Sharma et al., 2009, presented another review highlighting that there was scarce literature on the melt fraction studies of PCM used in the various applications for storage systems. Many of these applications have been studied widely in the last years; most are related to buildings and several to heat exchange between PCM and air as the heat transfer fluid:

- Ceiling cooling systems and floor heating systems including a PCM storage device were studied by authors like Turnpenny et al., 2001, and Yanbing & Yinping, 2003.
- Free-cooling has demonstrated to be an attractive application for latent heat storage using PCM. This application is reported in the work carried out by Butala & Stritih, 2009, and Lazaro et al., 2009a.
- Solar air heating systems are important in many industrial and agricultural applications, such as those reported in the papers by Kürklü, 1998.
- Other interesting possibilities are temperature maintenance/control in rooms with computers or electrical devices, and the pre-cooling of inlet air in a gas turbine (Bakenhus, 2000).

In any case, it is crucial to achieve efficient heat exchange between the heat transfer fluid and the PCM. This point is strongly affected by the heat exchanger geometry, as the TES unit has limited periods of time to solidify. Lazaro, 2009, compared the PCM-air heat exchange geometries studied by different researchers (Arkar et al., 2007; Turnpenny et al., 2000; Zalba et al., 2004; Zukowsky 2007). The author pointed out the difficulty of comparing between the different results provided by the authors, since each one show the results in its own way. Therefore, Lazaro concluded the need to standardize for proper comparison. Lazaro et al., 2009b, also presented experimental results for melting stage of real PCM-air heat exchangers pointing out the importance of the geometry. Geometry issues also affect the pressure drop of the TES unit and the air pumping requirements of the system, i.e., the electrical energy consumption. Regarding experimental studies, the evaluation of the thermal behaviour of the TES unit under statistical approaches or mathematical fitting leads to expressions that are very useful tools when designing such units. Among others, Butala & Stritih, 2009, and Lazaro et al., 2009b, followed this methodology when they evaluated their results.

In this chapter, a specific case study on slab geometry of a PCM-air heat exchanger is presented for temperature maintenance in rooms. However, the methodology posed here can be extrapolated to other different PCM geometries and system setups.

2. Pre-design: important factors

Since non-pure substances have lower costs than pure materials, they are used in commercial PCM. The characterization of the PCM and its encapsulation material are required to choose the optimal PCM and to design the heat exchanger for each application. The thermophysical properties of the PCM as a function of temperature are essential to the numerical model. Such information is not available for commercial PCM. This section therefore aims at the development of an adequate methodology to characterize PCM. Subsequently, the design of an experimental setup is explained, directed towards the determination of the enthalpy vs. temperature curves, by using the T-history method. The setup was built and a methodology was proposed to verify the T-history setups. The same methodology is applied to determine thermal conductivity, another essential thermal property regarding heat transfer. As a result of the application of the existing methods to analyze the liquid and solid phases, the most suitable method is chosen and the setup was started up. Besides the energy storage capacity and the thermal conductivity as a function of temperature, other properties are also important to be known, such as the compatibility of the PCM with the encapsulation material.

2.1 Determination of enthalpy as a function of temperature

In order to obtain the most suitable method to determine enthalpy as a function of temperature during the solid-liquid phase change, two main thermal analysis methods were studied: differential scanning calorimetry (DSC) and adiabatic calorimetry. In addition, a customized method was studied: the T-history method. A complex review of the work on thermophysical properties was carried out with some conclusions being (Lazaro, 2009):

- DSC is the most used method for determining the storage capacity because it is the most common commercial device (Zhang D. et al., 2007).
- There are several problems with using DSC for non pure and low thermal conductivity substances (Arkar & Medved, 2005).
- The number of authors that use the enthalpy vs. temperature curves to express the storage capacity of PCM is increasing (Zalba et al., 2003).

DSC, adiabatic calorimetry and T-history method were studied and compared. Factors considered in the method selection are: sample size, heating and cooling rate, obtainability of the h-T curve, introduction to the market, easiness to build, cost, use, maintenance. The T-history method was selected as it provides the enthalpy vs. temperature curves and also uses sample sizes and heating/cooling rates similar to those used in real applications.

2.1.1 The T-history method

Zhang et al., 1999, developed a method to analyze PCM enthalpy. The T-history method is based on an air enclosure where the temperature is constant and two samples are introduced at a different temperature from the temperature in the air enclosure. During cooling processes, three temperatures are registered: the ambient (air enclosure) and those of the two samples. The two samples are one reference substance whose thermal properties are known (frequently water) and one PCM whose thermal properties will be determined with the results of the test. Figure 1 shows the basic scheme of the T-history method.

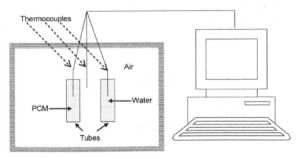

Fig. 1. Scheme of T-history installation.

The basic aspects of the T-history method are (Zhang et al., 1999):

- Heat transfer is one-dimensional in the radial direction since the samples containers are long cylinders.
- Containers with water and PCM samples are designed with Bi<0.1 and therefore are considered capacity systems.
- Heat transfer occurs by free convection between the samples and air. Containers must be identical in order to have a very low and almost the same free convection coefficient.

To evaluate the temperature vs. time evolution, Zhang proposed three stages: liquid, phase change, and solid. Therefore, with this method it is possible to obtain $c_{p,liquid}$, $c_{p,solid}$ and h_{sl}. Marin et al., 2003, made improvements, based on the finite increments method, in order to obtain the h-T curves. Figure 2 shows how the calculations were carried out.

$$m_p \Delta h_p (T_i) + m_t c_{pt}(T_i)(T_i - T_{i+1}) = hA_t \int_{t_i}^{t_i + \Delta t_i} (T - T_{\infty,a}) dt = hA_t A_i \tag{1a}$$

$$\left[m_t c_{pt}(T) + m_w c_{pw}(T) \right](T_i - T_{i+1}) = hA_t \int_{t'}^{t' + \Delta t'} (T - T_{\infty,a}) dt = hA_t A_i' \tag{1b}$$

$$\Delta h(T_i) = \left(\frac{m_w c_{pw}(T_i) + m_t c_{pt}(T_i)}{m_p} \right) \frac{A_i}{A_i'} \Delta T_i' \frac{m_t}{m_p} c_{pt}(T_i) \Delta T_i \tag{1c}$$

$$h_p(T) = \sum_{i=1}^{N} \Delta h_{pi} + h_{p0} \tag{1d}$$

$$c_p = \partial h / \partial T \tag{1e}$$

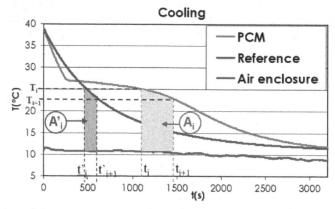

Fig. 2. Calculation of the improvements achieved by Marin et al., 2003.

The set of equations 1 summarize the calculations considering the improvements. There A_t denotes the tube lateral area, m_p the PCM mass, m_t the tube mass, c_{pt} the specific heat of the tube, c_{pw} the specific heat of water, h the convection coefficient whereas $h(T)$ denotes enthalpy. The little temperature steps, ΔT_i, varies in accord to the corresponding time intervals for the PCM (Δt_i=t$_{i+1}$-t$_i$) and for water ($\Delta t'_i$=t'$_{i+1}$-t'$_i$). The integral of the temperature difference against time, is the area under the curve in Figure 2 for the PCM (A_i) and for water (A_i').

2.1.2 Design of a new installation to implement the T-history method

When analyzing errors with T-history, the most important factor is the precision in the temperature measurement. Thermal sensors used in previous implementations have been thermocouples, while Pt-100 was chosen for this new installation due to the higher precision: ±0.05°C with a 4 threads assembly. However, Pt-100 has a longer response time, but will not affect the results provided that the response time is the same for all temperature measurements. This objective is achieved by using Pt-100 of the same manufacture set, and characteristics will be identical. Enthalpy is expressed in a mass unit basis; therefore the precision in mass measurements is as important as the precision in temperature measurements. A 0.1 mg precision scale is used to measure the mass of samples. The sample containers have been designed so that the method standards are fulfilled (Bi<0.1). Churchill-Chu (Marin & Monne, 1998) natural convection correlations for cylinders were used to calculate the suitable radius/length rate of the tubes. The chosen material was glass, since it allows the observation of the phase change process. Cylinders of 13 cm in length and 1 cm in diameter were used. A data logger was used with a RS-232 connection with 22 bits and 6 ½ resolution. A thermostatic bath (0.1 K precision) was used to fix the initial temperature of the samples. A calculation software, especially developed in Labview, was used to obtain the h-T curves. The new T-history implementation based its improvements on:

- Obtainability of the h-T curves during cooling and heating.
- Horizontal position of samples in the air enclosure, minimizing convective movements.
- Utilization of more precise instrumentation.
- A program designed (Labview) for calculations and real time view of the measurements.
- A guarantee that there is no contribution of heat transfer by radiation.

Examples of T-history analysis applied to two typical PCM (organic and inorganic) are shown in Figure 3. Typical phenomena as hysteresis or sub-cooling can also be observed.

The objective of analyzing organic and inorganic substances is to confirm the expected differences in behaviour: the inorganic PCM presents the sub-cooling phenomenon that occurs during cooling, presenting more hysteresis and quite higher stored energy density when compared to organic PCM.

The procedure used was: mass measurements of the samples and sample containers using a precision scale, then the Pt-100 were introduced into the samples (one into the PCM and one into the water), and the tubes were inserted into the thermostatic bath at the desired initial temperature. The initial temperature depends on the PCM to be tested as well as if it is for a heating or a cooling test. For a heating test, the initial temperature must be lower than the phase change temperature. For a cooling test, it must be higher. Once the temperature inside

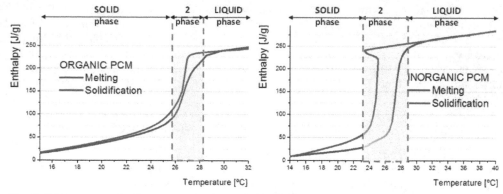

Fig. 3. T-history results for an organic PCM (left) and for an inorganic one (right).

the PCM and water is fixed, the tubes are inserted into the air enclosure and the measurement starts. The enthalpy was calculated as shown previously in equations 1.

Detailed information on the raw data and calculations can be found in Lazaro, 2009. An example of the outputs window (Labview application) of an arbitrary T-history test is shown in Figure 4.

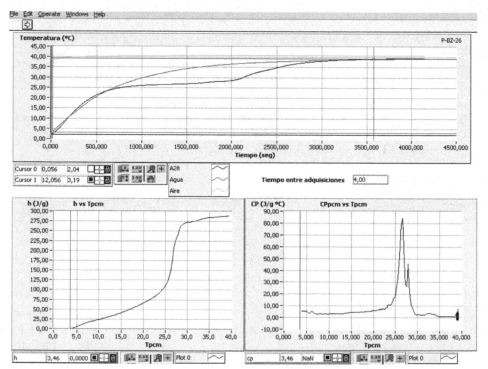

Fig. 4. Calculation outputs of a typical T-history test: measured temperatures (up), PCM enthalpy (down-left) and PCM specific heat (down-right).

2.2 Thermal conductivity

The same procedure to select the appropriate method to obtain the enthalpy vs. temperature curves was followed to find the method for thermal conductivity measurement in liquid and in solid phases. The most commonly used method is the hot wire method (Watanabe, 2002), nevertheless the temperature of the sample is measured with low accuracy and there is also the difficulty in measuring solid samples. A stationary parallel plate method (Mills et al., 2006) solves the problem of accuracy in temperature measurements, but in the liquid phase, convective movements affect the results. The Laser Flash is the only method that allows measuring the thermal diffusivity and sample temperature with accuracy, both in liquid and solid phases. It is based on a laser pulse that comes into contact with one surface of the sample and the temperature evolution on the opposite surface is measured by an infrared detector; therefore, the thickness of the sample must be perfectly determined. A mathematical evaluation of the temperature evolution allows the determination of the thermal diffusivity α of the sample (equation 2) and by measuring the heat capacity c_p with a DSC, also the thermal conductivity λ may be obtained:

$$\alpha = \frac{1.38}{\pi^2} \frac{L^2}{t_{1/2}}, \tag{2}$$

$$\lambda = \alpha \rho c_p, \tag{3}$$

where L is the sample thickness and $t_{1/2}$ is the time elapsed until half the temperature increment is achieved, and ρ is the density.

2.3 Other properties to consider

Although we have focused on enthalpy and thermal conductivity, there are other important properties and issues to consider such as: encapsulation compatibility (plastic-paraffin; salt hydrated-metal), toxicity, flammability, corrosion, thermal cycling, rheology, density, and volumetric expansion.

2.4 Geometry

The specific study system corresponds to a PCM-air heat exchanger acting as a TES unit. The unit is basically composed of PCM plates, the casing, and a fan that blows the air that circulates inside the equipment between the plates (see Figure 5). Although the set up could be arranged horizontally to reduce pressure drop and electrical consumption, the vertical distribution was a requirement because of the very first application (for temperature maintenance in telecom shelters, it should be a stand-alone system, hooked outside the façade, with the ability to plug in with a conventional chiller).

An important aspect in the design of PCM-air heat exchangers is the selection of an appropriate geometry of the PCM macroencapsulation. It is necessary to consider what will be the requirements that the storage system must satisfy and that will depend on the application. The heat transfer rate (absorbed or released), and the operation time, are two of the factors that generally will be considered. At least there are three typical options to select the shape of the macroencapsulation: plates, cylinders, and spheres. Here, plate shape is selected because it has been a deeply studied geometry since London & Seban, 1943. It involves: 1) Easy-to-control PCM thickness, which is a crucial design factor as it allows regulating elapsed times of the melting and solidification; 2) Uniformity of the PCM thickness and, therefore, of the phase change process; 3) Simplicity of the manufacturing process (both small scale and large scale) and versatility of handling (transportation, installation ...); 4) Commercial accessibility in a wide variety of plate-shaped encapsulations in different materials, both metallic and plastic.

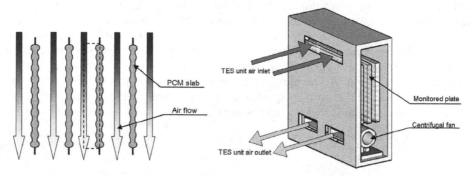

Fig. 5. PCM panels and air flow system (left); PCM-air heat exchanger (right).

Finally, the rigid metallic plate encapsulation has been selected to avoid both compatibility issues (Lazaro et al., 2006) as well as leakage problems detected previously (Lazaro, 2009) when using pouches.

2.5 Heat transfer mechanisms: basics

The basics of the heat transfer in PCM are compiled by Zalba et al., 2003, and discussed in a very understanding way by Mehling & Cabeza, 2008. The authors describe the basics of the heat transfer by means of: 1) Analytical models; 2) Numerical models; 3) Modelling; 4) Comparison of models vs. experimental; 5) Methods to improve the heat transfer.

3. Characterization of the heat exchanger

The main objectives of this section are:

1. How to test a prototype of PCM-air heat exchanger.
2. Gathering experimental results.
3. Analyzing data and obtaining empirical models.
4. Importance of the uncertainties in measurements and their propagation.

3.1 Experimental set up to test PCM-air heat exchangers

An experimental setup was designed to study different PCM-air heat exchangers (Dolado, 2011; Lazaro, 2009). A closed air loop setup was used to simulate indoor conditions. The setup design was based on the ANSI/ASHRAE STANDARD 94.1-2002 "Method of Testing Active Latent-Heat Storage Devices Based on Thermal Performance" (ANSI/ASHRAE, 2002). The setup is constituted of: 1) Inlet air conditioner allowing the simulation of different operating modes (5 kW air chiller and 4.4 kW electrical resistance); 2) Air flow measurements; 3) Difference between inlet and outlet air temperature measurements (thermopile); 4) Inlet and outlet air temperature and humidity measurements; 5) PCM and air channels temperature measurements (31 thermocouples); 6) Data logger and data screening; 7) Air ducts and gates; 8) PID controller.

The energy balance of air between the prototype's inlet and outlet is utilized to evaluate the cooling (equation 4). As the main parameters are the air flow and the air temperature difference between the inlet and the outlet, the accuracy depends on the precision when measuring these parameters. The methods used are:

$$\dot{Q} = \dot{m}_{\text{air through HX}} \cdot \Delta h_{\text{air}} \approx \dot{m}_{\text{air through HX}} \cdot c_{p_{\text{air}}} \cdot \Delta T \tag{4}$$

- Air temperature difference: thermopile. There were difficulties to overcome in this measurement: a long period of time with little temperature difference; the temperature distributions along the air ducts due of its dimensions; and accuracy, which is required since it is a main parameter of evaluation. A thermopile was chosen as it is recommended by the ANSI/ASHRAE standard to overcome those difficulties.
- Air flow: energy balance of electrical resistances. The air temperature changes during tests, therefore most of air flow measurement methods are not suitable for transitory measurements. Mass flows depend only on the fan velocity; therefore they are measured by applying an energy balance to the electrical resistances.
- Air humidity: 2 sensors were used to measure air humidity at the inlet and outlet. Latent energy variation was negligible in the air energy balance for cooling power evaluation.

The reader can find more information on the experimental setup in Lazaro, 2009.

3.2 Two prototypes

Two real-scale prototypes of PCM-air heat exchangers were constructed and incorporated into the experimental setup to characterize them. Initially tests were conducted with the equipment filled with bags of a hydrated salt PCM (prototype 1). Subsequently, the bags were replaced by plates of a paraffin based PCM, and the unit was tested filled with plates. These two geometries were arranged vertically and parallel to the airflow. The casing of the heat exchanger unit used in both cases was the same. PCM thickness was a critical parameter to obtain the required cooling rates (Dolado et al., 2007). Vertical position was a requirement; therefore a metallic grid was used to force PCM thickness below a maximum in vertical position. The experimental setup built to test this kind of heat exchangers is shown in Figure 6. Tests using a constant inlet air temperature setpoint were accomplished.

Figure 7 (left) shows the cooling power evolution in prototype 1. Results showed that the cooling rates were very low and the total melting times were double the melting design time (2h). Different air flow rates were tested. As it can be seen in figure 8 (left), the air flow influence on melting times and cooling rates were negligible (in the figure *HH:mm* denotes time, hours:minutes). Cooling power does not increase by a rise of air flow rates. Indicating that, contrary to what was at first designed, heat transfer by conduction inside the PCM resistance is dominant. The prototype was opened to confirm the diagnosis, and PCM

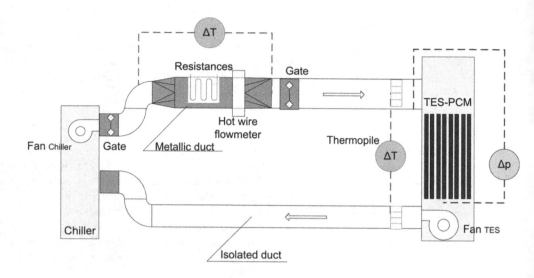

Fig. 6. Experimental installation arrangement to test PCM-air heat exchangers prototypes.

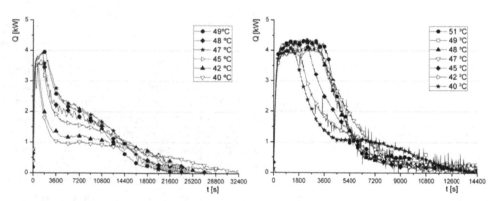

Fig. 7. Cooling rate evolution in prototype 1 during tests with different inlet air temperatures (left); in prototype 2 (right).

leakages were found out. Some pouches were torn and the metallic grid was deformed by the pushing force of the solidification process of the PCM inside pouches. PCM thickness was twice higher than the designed, causing a higher and dominant heat transfer resistance by conduction inside the PCM. Therefore melting times were higher and flow rate had almost no influence. This prototype did not fulfil melting time requirements and was discarded.

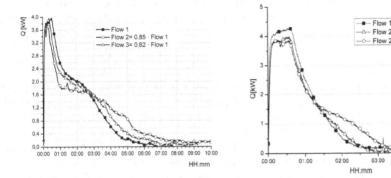

Fig. 8. Cooling rate evolution with constant inlet air temperature in prototype 1 during tests with different air flow rates (left); in prototype 2 (right); Flow 1 is the mass flow, equals to 0.34 kg/s.

Prototype 2 was designed using aluminium panels filled with organic PCM. Configuration was also vertical. Air flows were parallel to the panels from top to bottom. Due to the fact that the PCM in this prototype was organic, it presented lower thermal conductivity than PCM in prototype 1. Furthermore, the total stored energy in prototype 2 was also lower than in prototype 1. The first tests were accomplished using constant inlet air temperature. As it can be seen in figure 7 (right), cooling rates were higher than those obtained with prototype 1 and the melting times were half the melting times with prototype 1.

Different air flow rates in prototype 2 were tested. It was observed that it had influence on the melting time and cooling power. Figure 8 (right) shows that for the lowest air flow rate, the heat rate curve changes its shape and is more similar to prototype 1. This indicates that heat transfer by conduction inside the PCM starts to be relevant when compared to air convection. The first results of prototype 2 were satisfactory, so more test were planned to evaluate its behaviour under real conditions. Two types of experiments were accomplished: constant rise of inlet air temperature and constant heating power. Temperature rise ramps were then set into the resistances controller: results showed that faster the rise, higher cooling power and lower melting time. For constant power tests, different heating powers of electrical resistances were fixed. Results showed that prototype 2 was able to maintain a cooling capacity over 3 kW for approximately 1h 30′ or approximately 1 kW for more than 3 h. This result is useful to design the optimal operation mode depending on the application.

3.3 Experimental results

The total energy exchanged during melting and solidification, as well as the time elapsed until total melting/solidification were determined from the heat rate curves experimentally obtained. The influence of the inlet air temperature and air flow was studied, and results showed that the continuous thermal cycling of the unit is a repetitive process: running experiments with similar conditions led to the same thermal behaviour; no degradation in the PCM properties was noticed. Pressure drop was measured for different air flows. Depending on the inlet air temperature, full solidification of the PCM could be achieved in less than 3 h for an 8 °C temperature difference between the inlet air and the average phase change of the PCM. Average heat rates of up to 4.5 kW and 3.5 kW for 1 h were obtained for melting and solidification stages, respectively (Dolado et al., 2011b; Dolado, 2011; Lazaro, 2009; Lazaro et al., 2009a).

3.4 Empirical models

From experimental results, the empirical models were built aimed at simulating the thermal behaviour in the tested heat exchanger in different cases. These simulations were used to evaluate the technical viability of its application. Since the thermal properties of PCM vary with temperature, a PCM-air heat exchanger works as a transitory system and therefore, its design must be based on transitory analysis. This section shows that PCM selection criteria must include the power demand. The conclusions obtained for the PCM-air heat exchange can be useful for selecting PCM for other heat exchanger applications that use the tested geometry as well as for applications that use such technology: green housing, curing and drying processes, industrial plant production, HVAC, free-cooling.

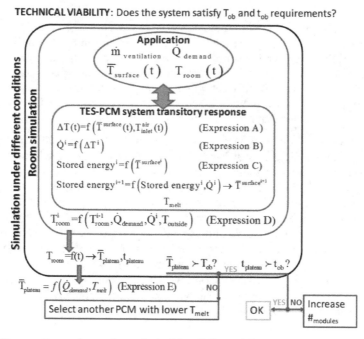

Fig. 9. Flow diagram to evaluate the technical feasibility of the system (Lazaro et al., 2009b).

The technical viability of TES with PCM systems depends on the capability to maintain the temperature below a maximum level (T_{ob}), during a specific period of time (Δt_{ob}). Figure 9 shows a flow diagram of the relevant parameters to test the technical viability of a system built with a specific geometry and an established air flow, for the heat exchange between air and PCM. To simulate a room under various conditions, the internal, external, and ventilation loads of the room, as well as the transitory response of the TES system using PCM must be known. Such transitory response was obtained by an empirical model built from all experimental outcomes under real conditions. The results to test prototype 2 with constant heating power were evaluated and an average temperature was obtained for the air temperature plateau ($\overline{T}_{plateau}$). There was obtained a linear correlation (equation 5) between the average plateau temperatures and the heating power of the electrical resistances ($\dot{Q}_{resistances}$). The origin ordinate was the average phase change temperature of the PCM used.

$$\overline{T}_{plateau} = 26.6 + 1.58\,\dot{Q}_{resistances} \tag{5}$$

$$\Delta T[K] = -1.4683 - 1.10943 \cdot \overline{T}^{surface}[^\circ C] + 1.10706 \cdot T^{air}_{inlet}[^\circ C] \tag{6}$$

The temperature on the surface of the PCM encapsulation was measured at 20 locations, distributed in such a way that the melting evolution can be studied and that any cold point can be detected. For each measurement, the average surface temperatures were obtained, as well as the plotting of the temperature difference between the inlet and the outlet (ΔT), the air temperature at the inlet of the TES unit ($T_{inlet}{}^{air}$) and the average surface temperatures ($\overline{T}^{surface}$). All locations come from experiments with the same airflow rate and heat exchange geometry. All measured locations were contained in a plane. A fitting tool was used to obtain the equation adjustment (equation 6). For the tested heat exchanger, expression A in figure 9 corresponds to equation 6. As it has been detailed, the heat exchanged with air was evaluated by an energy balance. The stored energy was then obtained for each measurement time step, as an accumulative result from the exchanged energy between air and PCM. The relationship between the stored energy and the average surface temperatures ($\overline{T}^{surface}$) corresponds to expression C in figure 9. Using the empirical model, different conditions were simulated, and the equations for the expressions shown in figure 9 were obtained, as well as various conclusions concerning PCM selection criteria (Lazaro et al., 2009b).

3.5 Uncertainties propagation

The guide EA-4/02 Expression of the Uncertainty of Measurement in Calibration, 1999, has been followed to estimate the uncertainty of measurements. Air flow was determined using an energy balance method that consists in applying an energy balance to the air flow that passes through the electrical resistances (equation 7). The air flow is measured with an accuracy of ± 0.026 kg/s (5.5% of measurement). Table 1 summarizes the expanded uncertainty estimation. The same procedure was followed to estimate the uncertainty of the cooling power determination. Equation 8 expresses the energy balance for the heat exchanger. Table 2 shows an example of uncertainty estimation in a cooling power measurement. In this case, the cooling power was measured with a ± 0.301 kW uncertainty (9%).

$$\dot{m}_{air} = \dot{Q}_{resistances} \Big/ \left(c_{p_{air}} \cdot \Delta T_{thermopile} \right) \tag{7}$$

$$\dot{Q}_{HX} = \dot{m}_{air}\, c_{p_{air}} \cdot \Delta T_{inlet\text{-}outlet} \tag{8}$$

	Expanded uncertainty	Standard uncertainty	Estimated value	Sensibility coefficient	Contribution to uncertainty
$\Delta T_{thermopile}[K]$	0.51	0.255	9.3	1	0.0008
Electrical power consumption of resistances [kW]	0.044	0.022	4.4	1	0.0000

Table 1. Air flow uncertainty of measurement estimation.

	Expanded uncertainty	Standard uncertainty	Estimated value	Sensibility coefficient	Contribution to uncertainty
$\Delta T_{thermopile}[K]$	0.51	0.255	7	1	0.0013
$\dot{m}_{air}$ [kg/s]	0.026	0.013	0.47	1	0.0008

	Sum of contributions	Estimated value	Standard uncertainty	Expanded uncertainty
$\dot{Q}_{HX}$ [kW]	0.089	3.29	0.151	0.301

Table 2. Example of cooling power uncertainty estimation.

The air temperature difference between the inlet and the outlet of the heat exchanger was measured using a thermopile and two Pt-100 in the centre of the air ducts. Measurements were compared during stationary periods in order to confirm the fact that a thermopile was more appropriate. Standard deviations of thermopile measurements are lower than the ones for Pt-100 differences and the mean values are all higher for the Pt-100 differences. This is due to the fact that Pt-100 are located at a specific point in the centre of the air duct whereas the thermopiles are distributed in the air duct cross surface.

4. Study of the heat transfer

In this section a theoretical model has been developed to perform the computer simulation of the thermal behaviour of a PCM-air heat exchanger, validating the theoretical model with the results obtained from the prototype in the experimental facility built for this purpose. In the archival literature, the approach of the solid-liquid phase change problem appears with different configurations, this section is focused on the case of macroencapsulated PCM, plate shape. Among the different numerical methods for solving the problem, in this section the energy equation is considered in terms of enthalpy as the governing equation can be applied at any stage, the temperature can be determined at each point, and therefore the values of the thermophysical properties can be evaluated. The PCM simulated are commercially available so the simulation involves, among other problems the non-linearity. The finite difference method for discretization of the governing equations is used. The models are based on 1D conduction analysis, using the thermo-physical data of the PCM measured in the laboratory. The models can take into account the hysteresis of the enthalpy curve and the convection inside the PCM, using effective conductivity when necessary.

4.1 Modelling the solid-liquid phase change

Modelling is a useful tool in a viability analysis of applications that involve TES by solid–liquid PCM. Therefore, there is a necessity to develop experimentally validated models that are rigorous and flexible to simulate heat exchangers of air and PCMs. When developing a model, the trade-off between rigour and computational cost is crucial. There are many options reported in scientific literature to face the mathematical problem of phase change as well as to solve specific particularities such as hysteresis phenomena (Bony & Citherlet, 2007) or sub-cooling (Günther et al., 2007). In the review by Zalba et al., 2003, the authors presented a comprehensive compilation of TES with PCM. The authors remarked that although there is a huge amount of published articles dealing with the heat transfer analysis of the phase change, the modelling of latent heat TES systems still remains a challenging task. When working with commercially available PCM (or mixtures or impure materials), the phase change takes place over a temperature range and therefore a two-phase zone appears between the solid and liquid phases. In these cases, it is appropriate to consider the energy equation in terms of enthalpy (Zukowsky, 2007b). When the advective movements within the liquid are negligible, the energy equation is expressed as follows:

$$\rho \cdot \partial h / \partial t = \nabla \left(\lambda \cdot \nabla T \right) \tag{9}$$

The solution of this equation requires knowledge of the h-T functional dependency and the λ-T curve. The advantage of this methodology is that the equation is applicable to every phase; the temperature is determined at each point and the value of the thermo-physical properties can then be evaluated. In thermal simulations of PCM, the accuracy of the results relies on the material properties' data (Arkar & Medved, 2005). In the geometry studied in this work, the main properties were enthalpy and thermal conductivity, but notice that the rate of melting/solidification can also depend on other material properties such as viscosity or density (Hamdan & Elwerr, 1996). In the models developed here the variation of thermophysical properties with temperature in all phases was considered.

4.2 Development of a 1D finite differences equations model for PCM plates

A PCM plate model was developed with finite differences, one-dimensional, implicit formulation. Implicit formulation was selected because of its unconditional stability. The basis model assumed only conduction heat transfer inside the PCM plate, in a normal direction to the air flow. The model analyzed the temperature of the airflow in a one-dimensional way. Due to its symmetry, the analyzed system was a division of the prototype. In the present work, the model was implemented in Matlab R2008b. The software implements direct methods, variants of Gaussian elimination, through the matrix division operators, which can be used to solve linear systems.

The study system is the PCM-air TES unit (figure 5, right). The air inlet was located on the upper side of the TES unit. The air flowed downwards in the TES unit, circulating parallel to the PCM slabs, exchanging energy with the PCM, and eventually was blown outside the TES unit by a centrifugal fan. The system was studied from the point of view of a single slab. As the PCM zone of the TES unit was insulated as well as due to the distribution of the slabs inside the TES unit, some symmetry relationships could be considered so only the dotted domain in figure 5 (left) was modelled. The nodal distribution of the mathematical model is shown in figure 10. Depending on whether the encapsulation is considered or not, two more nodes have to be included between the PCM surface and the airflow. In the experimental study, the heat transfer processes that take place inside the TES unit between the air flowing through the slabs and the PCM inside the slabs were: forced convection in the air, conduction in the shell of the aluminium slab, and conduction and natural convection in the PCM itself.

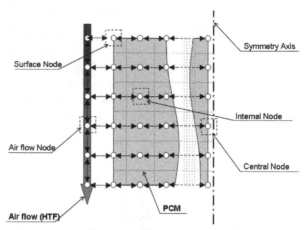

Fig. 10. Nodal distribution in the 1D plate system.

Nodes	Equations
Air Flow	$T_{air}=T_{air\text{-}1}-NTU_{air}\left(T_{air\text{-}1}-T_{surface}\right)$
PCM Surface	$T_{surface}=\left(T_{PCM}^{t\text{-}1}+FoT_{PCM+1}+FoBiT_{air}\right)\!/\!\left(1+Fo+FoBi\right)$
PCM Inner	$T_{PCM}=\left[T_{PCM}^{t\text{-}1}+Fo\left(T_{PCM\text{-}1}+T_{PCM+1}\right)\right]\!/\!\left(1+2Fo\right)$
PCM Central	$T_{PCM}=\left(T_{PCM}^{t\text{-}1}+2FoT_{PCM\text{-}1}\right)\!/\!\left(1+2Fo\right)$

Table 3. Node temperature equations not considering encapsulation.

Nodes	Equations
Air Flow	$T_{air}=T_{air\text{-}1}-NTU_{air}\left(T_{air\text{-}1}-T_{surface}\right)$
Air Surface	$T_{surface}=\dfrac{\left[T_{surface}^{t\text{-}1}+2\left(Fo_{enc}T_{enc}+Fo_{enc}Bi_{enc}T_{air}\right)\right]}{\left[1+2\left(Fo_{enc}Bi_{enc}+Fo_{enc}\right)\right]}$
Surface Encapsulation	$T_{enc}=\dfrac{\left[T_{enc}^{t\text{-}1}+2\left(Fo_{enc\text{-}PCM}T_{surface}+2Fo_{PCM\text{-}enc}T_{PCM}\right)\right]}{\left[1+2\left(Fo_{enc\text{-}PCM}+2Fo_{PCM\text{-}enc}\right)\right]}$
Encapsulation PCM	$T_{PCM}=\left[T_{PCM}^{t\text{-}1}+Fo\left(2T_{enc}+T_{PCM+1}\right)\right]\!/\!\left(1+3Fo\right)$
PCM Inner	$T_{PCM}=\left[T_{PCM}^{t\text{-}1}+Fo\left(T_{PCM\text{-}1}+T_{PCM+1}\right)\right]\!/\!\left(1+2Fo\right)$
PCM Central	$T_{PCM}=\left(T_{PCM}^{t\text{-}1}+2FoT_{PCM\text{-}1}\right)\!/\!\left(1+2Fo\right)$

Table 4. Node temperature equations considering encapsulation.

The dominant resistance of the process could be convection on the air side and not always conduction–convection in the PCM. In this case the thermal resistance of the encapsulation was very low, and therefore it was not necessary to consider encapsulation in the node system, as the heat transfer process was controlled by the convection on the air side and/or by the conduction–natural convection in the PCM. However, in other cases it is not always possible to disregard the thermal influence of the encapsulation, and therefore two models were developed: the first model did not take into account the thermal behaviour of the encapsulation and the second model did, and it was developed in order to be used for general purposes. The node equations of the two models are summarized in tables 3 and 4. Important aspects to consider when dealing with the simulation of this type of heat exchanger are as follows: friction factor (rugosity of the encapsulation surface), convection coefficient, thermophysical properties of the PCM (as functions of temperature), hysteresis, natural convection inside the PCM, thermal losses/gains through the TES casing, etc. More detailed information can be found in Dolado et al., 2011a, and Dolado, 2011.

4.3 Experimental validation: applying the uncertainties propagation approach to the model

The validation stage of a theoretical model has become a fundamental objective to evaluate the precision, accuracy and reliability of computer simulations used in design. Uncertainties can be associated with the own theoretical model, as well as with the measurement systems used to characterize the process of interest or even with the manufacturing process of the equipment. Therefore, assessing the validity of an approximation of a theoretical model must be carried out based on stochastic measurements to ensure the trust of designers in the use of the model. This improved knowledge of the theoretical model helps to know what are the most critical factors as model inputs and, therefore, indicates what should be more controlled in its determination or measurement. It also allows establishing an uncertainty band set around the solution bringing more rigor to the model simulations. The methodology of uncertainties propagation is an external method used to analyze the system by means of the input-output analysis, instead of the traditional equation of uncertainties propagation applied to a known function. The whole methodology followed in is summarized in the next steps: 1) To select the variables under study; 2) To allocate the probability distributions of each variable; 3) To generate samples for the different runs of the theoretical model (by means of Latin hypercube sampling); 4) To run the program once per sample; 5) To analyze the relations between the inputs and the outputs; 6) To classify the variables; 7) To determine the uncertainty of the theoretical model results.

For the current study, the following parameters that introduce uncertainty in the results, were considered and classified into three groups:

- Material properties (parameterized enthalpy-temperature curve, h-T, and thermal conductivity curve, λ-T);
- Air conditions at the inlet of the TES unit (temperature and airflow);
- Geometric parameters (PCM plate thickness and width of the air gap between plates).

The parameterization and the range of uncertainty assumed for all these parameters are detailed in Dolado, 2011. The confidence level is 97.5%. Furthermore, as the probability distribution of the different parameters is unknown, a normal distribution is taken in all cases. The study of uncertainties propagation is performed by numerical simulation of sets of input values of those parameters. Traditionally the random sampling technique has been used, which was followed by an improved version as the stratified sampling and subsequently the Latin hypercube (McKay et al., 1979) which is a valid technique used here. The variation range of the studied variable was analyzed. This result provides an estimate of the uncertainty range that would result as the TES unit is designed. This analysis sets the interval for the output variable of interest. For every particular time of the simulation the error was calculated as: 1) Histogram to the corresponding time of the results set of all simulations (the difference between the value in the current simulation and the reference case); 2) Function of the cumulative probability distribution; 3) Error in the given time at a certain confidence level; 4) Graphical representation of the reference case and of the error.

Parameter	b	ΔT	h_l	T_{sl}
Related with	Slope of the sensible heat	Thermal window	Latent heat	Average temperature of phase change
Reference value	3 kJ/(kg·K)	0.9 °C	170 kJ/kg	27 °C
Uncertainty	± 15 %	± 0.2 °C	± 20 kJ/kg	± 1 °C
Effect on h-T curve				

Table 5. Parameters setting of the enthalpy-temperature curve.

Variable	Reference value	Uncertainty
$\dot{V}$	1400 m³/h	± 86 m³/h
T	experimental curve of $T_{air,in}$, °C	± 0.6 °C

Table 6. Parameters setting of the inlet air.

The corresponding value for a cumulative relative frequency of 97.5% was taken as the uncertainty range in this approach. Running the theoretical model with the reference case conditions (tables 5 and 6) yielded to the results shown in figure 11. The figure shows that the heat rate has an initial plateau of about 4500 W, with a duration of 40 minutes, which reduces while reaching the complete melting of the PCM. The full melting takes place two hours after the start of the process. From these graphical results, the responses of interest were obtained. Among the responses of interest provided by the theoretical model, the analysis was focused on the following responses: average heat rate in the first hour ($\dot{Q}_{average,1h}$), and time until the air reaches 32 °C at the outlet ($t_{Tair,out=32°C}$), as both responses pose a greater interest in practical operation of the TES unit. Immediately afterwards, applying the Latin hypercube sampling to run the numerical simulations, instead of getting a single result of $\dot{Q}_{average,1h}$, a complete distribution was obtained. Figure 11 shows the evolution of the exchanged heat rate in the melting process with the associated uncertainty interval (97.5%) and also the results of the relative error in heat rate. It is observed that the relative error is below 10 % until the process is approaching the end of the melting stage. Is then when the absolute values of heat rate are smaller and the relative error increases until the melting ends, as expected. This result is analogous in the solidification stage. As stated by Dolado et al., 2011b, and Lazaro et al., 2009a, considering the instrumentation used in the experimental setup, an uncertainty of 9 % in terms of the heat rate during the first hour of a typical test process is obtained. The calculations to determine the uncertainty in the measurement of the heat rate are shown in table 7. The uncertainty is estimated for each measurement so that, according to the EA Guide 4/02 Expression of the Uncertainty of Measurement in Calibration, 1999, a band of uncertainty associated with the experimental heat rate curve can be determined (figure 12).

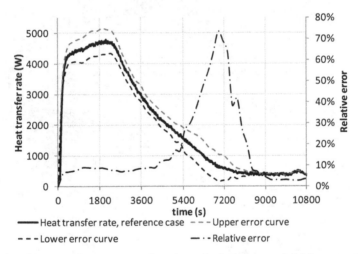

Fig. 11. Simulated heat rate, uncertainty bands and relative error (melting stage).

	Expanded uncertainty	Standard uncertainty	Estimated value	Sensibility coefficient	Contribution to uncertainty
$\Delta T_{thermopile}[K]$	0.51	0.255	ΔT_i	1	$(0.255/\Delta T_i)^2$
$\dot{m}$ [kg/s]	0.026	0.013	0.36	1	0.0013
c_P [J/(kg·K)]	2	1	1007	1	$9.86 \cdot 10^{-7}$
	Sum of contributions		Estimated value	Standard uncertainty	Expanded uncertainty
$\dot{Q}$ [W]	SQRT [(0.0013+9.86·10^{-7}+(0.255/ΔT_i)^2]		$\dot{Q}_i$	$w_{\dot{Q}} \cdot \dot{Q}_i$	$2 \cdot w_{\dot{Q}} \cdot \dot{Q}_i$

Table 7. Uncertainties determination of the experimental heat rate.

As expected, the relative errors grow as the absolute value of the heat rate decreases (figures 11 and 12). This is because the expanded uncertainty associated with the measure of the thermopile is ± 0.51 °C (Lazaro, 2009) so that the error increases as the temperature difference between the air at the inlet and at the outlet of the TES unit decreases. Figure 12 shows the overlap between the experimental curve (including the lower and upper limits associated with its uncertainty) and the simulation (including the uncertainty of the response heat rate calculated applying the reported technique). The agreement is significant in most of the process, finding the more relevant discrepancies as the curve reaches the end of the corresponding stage of the cycle (i.e. as the heat rate values are smaller).

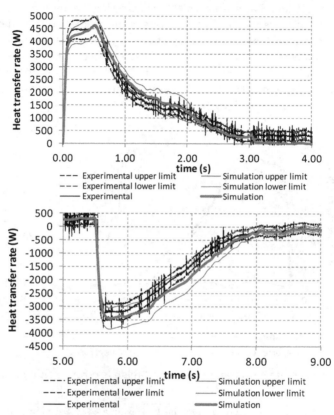

Fig. 12. Comparison of experimental and simulated results (including their corresponding uncertainty bands) for the melting (up) and solidification (down) of a thermal cycle.

5. Design focused on applications. Feasibility

This section will describe how to design the heat exchanger once an application is specified. Free-cooling and temperature maintenance in rooms with special requirements possess high potential for PCM application in different countries according to their climate. Until now, the low thermal conductivity of PCM and air hindered the development of suitable heat exchangers. This section has as the overall objective to apply methodologies to study PCM and PCM-air heat exchangers that allow the development of applications with technical and economical viability. Finally, using the combined technique of design of experiments (hereafter DOE) and simulations, the feasibility of the possible application of this type of equipment is studied for temperature maintenance in rooms. Because the simulation itself is not a design tool, this methodology is proposed to size the equipment. This technique greatly reduces the time spent in performing the simulations required to find the optimal equipment (Del Coz Díaz et al., 2010) as well as and a potential cost saving on the experimental (Del Coz Díaz et al., 2010; Gunasegaram et al., 2009) if the prototype-model similarity relations are met. Moreover, contrary to a sequential analysis, it is reasonable to use a mathematical and statistical methodology that allows planning the sequence of experiments on the philosophy of maximum information with minimum effort.

5.1 Empirical model: simulations of a case study and modular design

An empirical model was built from the experimental results described in the previous section. The aim was to simulate the thermal behaviour of the tested heat exchanger in different cases. These simulations were used to evaluate the technical viability of application. The model describes the temperature evolution of a room with an internal cooling demand ($\dot{Q}_{demand}$), where the PCM-air heat exchanger is operating and there is a ventilation system. The enclosure temperature was considered to be the average between the outside temperature and the room temperature. A diagram of the room is shown in figure 13. Expression D in figure 9 is equivalent to equation 9, expressing the energy balance applied to the air inside the room.

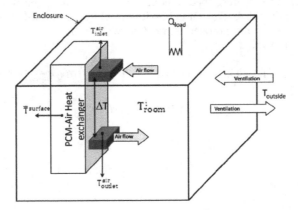

Fig. 13. Schematic diagram of the room in which the temperature is evaluated.

$$m\, c_{p,enclosure} \cdot \left[\left(\frac{T_{room}^i + T_{outside}}{2} \right) - \left(\frac{T_{room}^{i-1} + T_{outside}}{2} \right) \right] = \rho_{air} \cdot V \, c_{p,air} \cdot \left(T_{room}^i - T_{room}^{i-1} \right)$$

$$= \left[\dot{m}_{ventilation}\, c_{p,air} \cdot \left(T_{outside} - T_{room}^i \right) + \dot{Q}_{demand} - \dot{m}_{air\,HX}\, c_{p,air}\, \Delta T^{i-1} \right] \cdot \Delta t(i,i\text{-}1)$$

(9)

where ΔT is obtained at each instant as a function of $\overline{T}^{surface}$ and the inlet air temperature, $T_{inlet}{}^{air}$ (at instant i equal to $T_{room}{}^i$); and the $\overline{T}^{surface}$ at instant i is obtained from the stored energy evolution.

The real-scale PCM-air heat exchanger tested was constituted of 18 parallel modules (#$_{modules}$ denotes de number of PCM modules in the heat exchanger). A module is constituted by a metallic PCM container between two air channels. The pressure drop is the same for each module, and the air distribution through the air channels can be considered uniform. The unitary air flow through a module is the mass air flow ($\dot{m}_{air\ HX}$) divided by 18. Since the geometry and the air flow were maintained identical, the total stored energy for one module (E_t^{mod}) between two temperatures is the stored energy for the real-scale PCM-air heat exchanger between the two temperatures divided by 18 (equation 10). The total melting time depends on E_t^{mod} and on the cooling power demand (equation 11).

$$\text{Stored energy}=\#_{modules}\cdot E_t^{mod} \tag{10}$$

$$t_{melt}=\text{Stored energy}/\dot{Q}_{demand}=\left(\#_{modules}\cdot E_t^{mod}\right)/\dot{Q}_{demand} \tag{11}$$

$$\overline{T}_{plateau}=T_{melt}+1.58\ \dot{Q}_{resistances}=T_{melt}+1.58\ \dot{Q}_{demand}\cdot 18/\#_{modules} \tag{12}$$

The 1.58 value in equation 13 comes from the linear correlation between the average plateau temperatures and the heating power ($\dot{Q}_{resistances}$) data obtained experimentally. The origin ordinate is the average phase change temperature of the PCM used. The relationship between the average phase change temperature (T_{melt}) and the cooling power demand (expression E in figure 9) is described in equation 12. Assuming that the origin ordinate in the adjustment equation 13 is T_{melt} , it is possible to define the number of modules and the T_{melt} needed for a given cooling power demand, as well as the T_{ob} and Δt_{ob} to maintain such a level (equations 14 and 15).

$$\Delta T[K]=-1.4683-1.10943\cdot\overline{T}^{surface}[^{\circ}C]+1.10706\cdot T_{inlet}^{air}[^{\circ}C] \tag{13}$$

$$T_{melt}=T_{ob}-1.58\ \dot{Q}_{demand}\cdot 18/\#_{modules} \tag{14}$$

$$\#_{modules}=\dot{Q}_{demand}\cdot \Delta t_{ob}/E_t^{mod} \tag{15}$$

For example, in the case where a 2 kW cooling power demand is required and a temperature level of 25 °C maintained during 2 h using a TES system, then 18 heat exchanger modules filled with a PCM of the same thermal properties of the one used in prototype 2 but with a T_{melt} of 21.8 °C would be needed. The same case with a cooling power of 4 kW would require a T_{melt} of 18.7 °C (see table 8).

$\dot{Q}_{demand}$ [kW]	T_{ob} [°C]	t_{ob} [s]	T_{melt} [°C]	$\#_{modules}$
2	25	7200	21.8	18
4	25	7200	18.7	18

Table 8. Design conclusions for different cooling demands.

5.2 Theoretical model: DOE applied to simulations, improving design

The empirical model can give a very fast approach of relevant design parameters such as the PCM average phase change temperature. However, if we want to analyze the behaviour of the equipment when modifying any other parameter or variable, or if we need to improve/optimize the design, we have to move to the numerical model.

As a starting point we will continue using the case brought by Lazaro, 2009, which provides that, for proper running of the electronic equipment, the maximum air temperature in the room should be between 38 °C and 48 °C, in particular we will establish it at 44 °C. The heat generation of the electronic equipment is 5 kW. For the evolution of temperature inside the room, an energy balance was stated with the following simplifications: 1) the cooling effect of the terrain was not considered. The ground floor area is supposed to be occupied by the equipment; 2) exterior ventilation is introduced only when it is favourable, and considering that the environment outside the house is 40 ° C (worst case).

The idea behind this system is that after a failure of the conventional cooling system, the TES unit is intended to smooth the evolution of the temperature of the room so that it extends the time to reach a certain threshold temperature value. The aim is this period to be about two hours, so technicians have sufficient time to reach the place where the room is located and to repair the damage of the cooling system without having to stop the electronic equipment. A series of restrictions put on the TES system follow:

- Dimensions limitation due to the telecommunications shelter: the maximum length of the system is limited to 2.5 m (height of the shelter) which limits the section of the PCM to 1.25 m. Likewise, the width of the unit is also limited to 5 m due to the wall;
- Electrical power consumption limitation of the fan, so it can be supplied by batteries without being essential a connection to the grid. Pressure drop should be less than 30 Pa.

M_{PCM} [kg]	$\dot{V}$ [m³/h]	e_{plate} [mm]	e_{air} [mm]	Finishing
132	1340	6.5	12	3

Table 9. Operating conditions.

The operating conditions are shown in table 9 and the simulation results with the theoretical model of the unit proposed by Lazaro, 2009, are shown in figure 14.

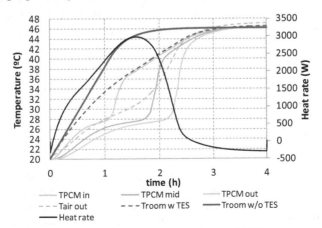

Fig. 14. Theoretical model simulated results of Lazaro's case (2009).

As it can be seen in the results of the simulation the contribution of the storage equipment is remarkable: the time spent to reach the room 38 °C is 1h 40min (determined by the red dotted line), extending almost 40 minutes than if there was no storage system (red line). Table 10 compiles the main results.

%Melt	Investment [€]	$t_{aditional, T=38°C}$ [min]	$t_{aditional, T=44°C}$ [min]	Δp [Pa]
69.47	3924	36	61	36

Table 10. Main results of the simulation with Lazaro's case (2009).

Factors	Domain	
	Level (-)	Level (+)
M_{PCM} [kg]	100	200
$\dot{V}$ [m³/h]	1000	2000
e_{plate} [mm]	6	14
e_{air} [mm]	10	20
Finish	1.5	2

Table 11. List of factors and their corresponding domain.

For the implementation of DOE the following factors and responses were considered:

- Factors (listed in table 11 along with their domain): mass of PCM, air flow, air channel width, thickness of the PCM plate, finishing of the plates (related to rugosity or to the presence of bulges in the surface of the plates).
- Responses: melting ratio in 3 hours, additional time for the air to reach a temperature of 38 ° C (compared with the evolution of temperature without unit TES) in the room, additional time for the air to reach a temperature of 44 ° C (compared with the evolution of temperature without TES unit) in the room, pressure drop, initial investment (mainly depending on the amount of PCM, the installed fan, the casing, and whether or not the plates have bulges on its surface).

5.2.1 Response optimization

Given that the main objective of the TES unit is to extend the time period during which the room temperature is below a certain temperature limit (in order to safeguard electronic equipment), the highest importance has set to that response. Table 12 lists the input parameters in the optimization. It has been considered that the most important requirement is to get the unit to extend as much as possible the time to reach the temperature limit of the air in the room, assigning the greatest importance to the maximum temperature limit (44 °C), $t_{aditional, T=44°C}$, and considering also important, but lesser, the time to reach the first temperature limit (38 °C), $t_{aditional, T=38°C}$, as well as the pressure drop, Δp (in order to be as lower as possible so that the electrical power consumption of the corresponding fan will be reduced). Also the investment and the melting ratio, %Melt, are interesting responses considered in the study, as they are related to economical and technical feasibility respectively. Once the objectives are defined, each variable is assigned a weight (between 0.1 and 10) and an importance (also between 0.1 and 10).

In this approach to the optimization, each of the values of the responses is transformed using a desirability function. The weight defines the shape of this function for each response and is related to the emphasis on achieving the target:

- A value greater than one emphasizes the importance of achieving the goal;
- A unit value gives equal importance to the objective and the limits;
- A value less than one puts less emphasis on the goal.

After calculating the desirability for each response, the desirability composite is calculated (weighted geometric mean of the single ones) that allows to obtain the optimal solution.

In this case, the same weight is set to each of the answers assuming a unit value. This will set the target as important as any value within the limits for the corresponding answer.

On the other hand, assigning a value to the importance of each answer is related to the importance given to each of the answers, and if any of these responses is more important than the others (the most important is a 10, the less important is a 0.1). The optimization results are shown in figure 15.

Response variable	Objective	Weight	Importance
$t_{aditional, T=44°C}$	Maximize	1	10
Δp	minimize	1	5
$t_{aditional, T=38°C}$	Maximize	1	5
Investment	minimize	1	1
%Melt	Maximize	1	1

Table 12. Optimization parameters.

What is interesting of the optimized results is the value of composite desirability as well as its trend according to each of the factors considered. The composite desirability obtained in this case (0.919) indicates that the values determined by the optimization nearly fulfil the requirements of the response variables. The trends of composite desirability for each factor allow to adjust their value (usually due to physical or technological constraints) while keeping high desirability values. However, at least there are two drawbacks to use this configuration: first, it does not respect the width limitation (this unit has a width of more than 10 meters), and secondly, when manufacturing the TES unit it will be more feasible to use a PCM thickness higher than 0.5 mm (proposed in the optimization). Thus, moving in the optimization plot to a greater value of PCM thickness without reducing too much the composed desirability and rounding parameters, a value of 2.5 mm in thickness is selected (which also meets the width restriction). Table 13 shows the results of the corresponding simulation. The results of the last proposed unit are somewhat unfavourable compared to the optimized unit, but the proposed thickness of PCM is much more realistic than the optimized one. Yet the responses provided by the proposed unit represent a storage that improves the very first one. The comparison of these results against the ones of the initial unit reflected that: a) Time to reach the target temperature of 44 °C increases: from 61 minutes it extends to 73 (19.7% improvement), being this a fundamental aspect of the application; b) The initial investment is reduced by 11%: from 3924 € to 3489 €; c) The PCM melting ratio is improved 23.2%; d) However, the volume occupied by the unit increments from 1.2 m³ to 3.8 m³.

Unit	%Melt	Investment [€]	$t_{aditional,\ T=38°C}$ [min]	$t_{aditional,\ T=44°C}$ [min]	Δp [Pa]
Proposed	92.64	3489	37	73	5
Optimized	100	3234	60	96	3

Table 13. Main results of the proposed and optimized units.

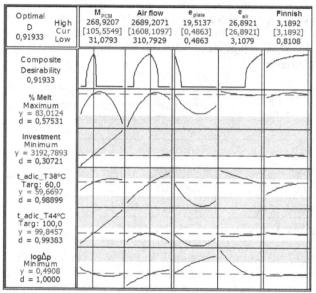

Fig. 15. Optimization plot results.

5.2.2 Model-prototype similarity

Dimensional analysis of these units show that the natural convection within the PCM is not going to be significant in any of the 2 units, being the heat transfer process by pure conduction for the second unit, and the ratio λ_{eff}/λ within the range of experimental validity for the other one (Dolado, 2011).Furthermore, since both Re and Bi numbers and NTU are within the range of experimental validity, the units can be used for design purposes.

5.3 Other applications

Keeping the temperature range, this type of heat exchanger can be applied in other different situations such as free-cooling, heat pumps, absorption solar cooling systems, greenhouses.

In any case, the DOE methodology proposed above could be followed to design a proper TES unit to the corresponding application (Dolado, 2011).

6. Conclusion

Methods to obtain enthalpy as well as the curves of thermal conductivity in the solid and liquid phases vs. temperature were proposed as a result of a critical analysis of the existing methods. A setup based on the T-history method was designed and built with significant improvements: 1) The possibility of measuring, for both organic and inorganic materials, cooling processes, therefore hysteresis and sub-cooling can also be studied; 2) The horizontal position decreases the error on enthalpy values since the liquid phase movements are minimized; 3) A Labview application allows the h-T curves to be directly obtained.

Results show that a heat exchanger using a PCM with lower thermal conductivity and lower total stored energy, but adequately designed, has higher cooling power and can be applied for free-cooling. Pressure drop is a key factor when designing any type of heat exchanger as it will determine the electrical energy consumption of the device. In the PCM-air heat exchangers with plates studied here, the pressure drop is ranged from 5 to 25 Pa. The analysis of the experimental data gathered accomplishes two aims: to develop empirical models of the TES unit and to come to a series of rules of thumb. Both are useful tools to design such kind of heat exchangers. For total energy storage strategy, the duration time of the cooling capacity of PCM heat exchanger depends on the cooling power demand. To validate the theoretical model developed, an uncertainties propagation analysis is proposed; here, the difference between the experimental and the simulation is less than 10% in terms of heat rate. The combined methodology of Design of Experiments applied to the numerical simulations seems to be a valid tool for design this kind of heat exchangers. When applied to the case study of temperature maintenance in a room, time to reach the maximum air temperature in the room was increased (19.7%), the initial investment was reduced by 11% and the PCM melting ratio was improved by 23.2%, as a drawback, the volume occupied by the unit was increased around 3 times.

7. Acknowledgment

The authors would like to thank the Spanish Government for the partial funding of this work within the framework of research projects ENE2005-08256-C02-02 and ENE2008-06687-C02-02. Pablo Dolado would specially like to thank the former Spanish Ministry of Education and Science for his FPI grant associated with the research project. The authors also wish to thank the company CIAT for the support given in the early stages of the experimental work. Special thanks are extended to Mr. Miguel Zamora, CIAT R&D Manager, for his collaboration.

8. Nomenclature

A	$[m^2]$	heat exchange area
A_t	$[m^2]$	tube lateral area
A_i	$[m^2]$	area under the T-t curve, for the PCM
A_i'	$[m^2]$	area under the T-t curve, for water
b	$[J/(g \cdot K)]$	parameter associated with the slope of the curve in all-liquid phase and all-solid phase, sensible heat, heat capacity

$$C_{air} = \rho_{air} \cdot \frac{\dot{V}_{air}}{2 \cdot N_{walls}} \cdot Cp_{air} \ [J/(s \cdot K)] \qquad \text{heat capacity}$$

c_p	$[J/(kg \cdot K)]$	effective specific heat
$c_{p,\ air}$	$[J/(kg \cdot K)]$	specific heat of air
$c_{p,\ liquid}$	$[J/(kg \cdot K)]$	PCM effective specific heat in liquid phase
$c_{p,\ solid}$	$[J/(kg \cdot K)]$	PCM effective specific heat in solid phase
c_{pt}	$[J/(kg \cdot K)]$	specific heat of the tube
c_{pw}	$[J/(kg \cdot K)]$	specific heat of water
d		desirability parameter (ranges from 0 to 1)
e	[m]	thickness
e_{plate}	[m]	thickness of the PCM plate
e_{air}	[m]	thickness of the air gap between two PCM plates
E_t^{mod}	[kJ]	total stored thermal energy for one module
h	$[W/(m^2 \cdot K)]$	convection coefficient (when standing alone)
h	[J/g]	enthalpy
h_{sl}	[J/g]	PCM solid-liquid phase change enthalpy
h_l	[J/g]	enthalpy value in liquid phase, just after finishing the solid-liquid phase change
h_s	[J/g]	enthalpy value in solid phase, just before the start of the solid-liquid phase change
L	[m]	sample thickness
m_p	[kg]	PCM mass
m_t	[kg]	tube mass
$\dot{m}$	[kg/s]	mass flow
$\dot{m}_{air\ through\ HX}$	[kg/s]	air mass flowing through the heat exchanger
$\dot{m}_{ventilation}$	[kg/s]	ventilation air mass flow
M_{PCM}	[kg]	PCM mass
N		number of elements
$NTU_{air} = (h \cdot \Delta x \cdot w)/C_{air}$		number of transfer units
$\dot{Q}$	[W]	thermal power, heat transfer rate
$\dot{Q}_{average,\ 1h}$	[W]	average heat transfer rate in the first hour of storage unit operation
$\dot{Q}_{demand}$	[kW]	internal cooling demand
$\dot{Q}_{HX}$	[kW]	heat transfer rate in the heat exchanger
$\dot{Q}_{resistances}$	[kW]	heating power of the electrical resistances used in the experimental setup
t	[s]	time
t_{melt}	[s]	total melting time
$t_{additional,\ T=38°C}$	[s]	elapsed time to reach 38°C in the room
$t_{additional,\ T=44°C}$	[s]	elapsed time to reach 44°C in the room
	[s]	time until the air reaches 32°C at the outlet of the storage unit
$t_{1/2}$	[s]	time elapsed until half the temperature increment is achieved
T	[K, °C]	temperature

$\bar{T}_{surface}$	[°C]	average surface temperature of the PCM
$\bar{T}_{plateau}$	[°C]	average of air temperature during the plateau, obtained either from the evolution of room temperature when it is simulated or from the air temperature at the heat exchanger outlet when it is measured on the experimental setup
T_{room}^{i}	[°C]	room temperature at i instant
$T_{outside}$	[°C]	outdoors air temperature
T_{melt}	[°C]	average PCM melting temperature
T_{ob}	[°C]	air temperature plateau objective
	[°C]	air temperature at the inlet of the storage unit
T_{sl}	[°C]	average phase change temperature of PCM
$\dot{V}$	[m³/h]	volumetric flow
w		uncertainty contribution
#$_{modules}$		number of PCM modules in the storage unit
%Melt		ratio of PCM melted, percentage

Greek symbols:

α	[m²/s]	thermal diffusivity
λ	[W/(m·K)]	thermal conductivity
$λ_{eff}$	[W/(m·K)]	effective thermal conductivity
ρ	[kg/m³]	density
Δh	[J/g]	enthalpy difference
Δp	[Pa]	pressure difference
ΔT	[K]	temperature difference
$ΔT_i$	[K]	temperature step
$ΔT_{thermopile}$	[K]	temperature difference of air between the inlet and outlet, measured using a thermopile
$Δt_i{=}t_{i+1}{-}t_i$	[s]	time interval for the PCM
$Δt'_i{=}t'_{i+1}{-}t'_i$	[s]	time interval for water
$Δt_{ob}$	[s]	plateau time objective
Δx, Δy	[m]	node length and height respectively

Acronyms and definitions:

Bi	Biot number
Fo	Fourier number
Re	Reynolds number

$$Fo_{enc} = \frac{\lambda_{enc} \cdot \Delta t}{\rho_{enc} \cdot Cp_{enc} \cdot e^2}$$

$$Fo_{PCM} = \frac{\lambda_{PCM}(T) \cdot \Delta t}{\rho_{PCM}(T) \cdot Cp_{PCM}(T) \cdot \Delta y^2}$$

$$Fo_{PCM-enc} = \frac{\lambda_{PCM} \cdot \Delta t}{(\rho_{enc} \cdot Cp_{enc} \cdot e + \rho_{PCM} \cdot Cp_{PCM} \cdot \Delta y) \cdot \Delta y}$$

$$Fo_{enc-PCM} = \frac{\lambda_{enc} \cdot \Delta t}{(\rho_{enc} \cdot Cp_{enc} \cdot e + \rho_{PCM} \cdot Cp_{PCM} \cdot \Delta y) \cdot e}$$

$Bi_{enc} = h_{air} \cdot e / \lambda_{enc}$

DOE	Design of Experiments
DSC	Differential Scanning Calorimetry
HTF	Heat Transfer Fluid
HVAC	Heating, Ventilation, and Air Conditioning
PCM	Phase Change Material
PID	Proportional Integral Derivative
TES	Thermal Energy Storage
1D	One Dimensional

9. References

ANSI/ASHRAE STANDARD 94.1-2002. (2006). *Method of Testing Active Latent-Heat Storage Devices Based on Thermal Performance* (ANSI approved), recently replaced by ASHRAE 94.1-2010.

Arkar, C. & Medved, S. (2005). Influence of accuracy of thermal property data of a phase change material on the result of a numerical model of a packed bed latent heat storage with spheres. *Thermochim Acta*, (Aug 2005), Vol. 438, No. 1–2, pp. 192–201, 0040-6031.

Arkar, C.; Vidrih, B. & Medved, S. (2007). Efficiency of free cooling using latent heat storage integrated into the ventilation system of a low energy building. *Int J Refrig*, (Jan 2007), Vol. 30, No. 1, pp. 134-143, 0140-7007.

Bakenhus, B.H. (2000). Ice storage project. *ASHRAE J*, (May 2000), Vol. 42, No. 5, pp. 64-66, 0001-2491.

Bony, J. & Citherlet, S. (2007). Numerical model and experimental validation of heat storage with phase change materials. *Energy Build*, (Oct 2006), Vol. 39, No. 10, pp. 1065–1072, 0378-7788.

Butala, V. & Stritih, U. (2009). Experimental investigation of PCM cold storage. *Energy Build*, (Mar 2009), Vol. 41, No. 3, pp. 354-359, 0378-7788.

Del Coz Díaz, J.J.; García Nieto, P.J.; Lozano Martínez-Luengas, A. & Suárez Sierra, J.L. (2010). A study of the collapse of a WWII communications antenna using numerical simulations based on design of experiments by FEM. *Eng Struct*, (Jul 2010), Vol. 32, No. 7,pp. 1792-1800, 0141-0296.

Dolado, P.; Lazaro, A.; Zalba, B. & Marín, J.M. (2007). Numerical simulation of heat transfer in phase change materials (PCM) for building applications. *Proceedings of Heat transfer in components and systems for sustainable energy technologies*, 2-9502555-3-1, Chambery, France, April 2007.

Dolado, P.; Lazaro, A.; Marin, J.M. & Zalba, B. (2011a). Characterization of melting and solidification in a real-scale PCM-air heat exchanger: Numerical model and experimental validation. *Energy Conv Manag*, (Nov 2010), Vol. 52, pp. 1890-1907, 0196-8904.

Dolado, P.; Lazaro, A.; Marin, J.M. & Zalba, B. (2011b). Characterization of melting and solidification in a real-scale PCM-air heat exchanger: Experimental results and empirical model. *Renew Energy*, (Apr 2011), Vol. 36, pp. 2906-2917, 0960-1481.

Dolado, P. (2011). *Thermal Energy Storage with phase change. Design and modelling of storage equipment to exchange heat with air*. Thesis, University of Zaragoza, 978-84-694-6103-7, Zaragoza, Spain. Access by (in Spanish): http://zaguan.unizar.es/record/6153

EA-4/02. (1999). *Expression of the Uncertainty of Measurement in Calibration*. European co-operation for Acreditation.

Gunasegaram, D.R.; Farnsworth, D.J. & Nguyen, T.T. (2009). Identification of critical factors affecting shrinkage porosity in permanent mold casting using numerical simulations based on design of experiments. *J Mater Process Technol*, (Feb 2009), Vol. 209, No. 3, pp. 1209–1219, 0924-0136.

Günther, E.; Mehling, H. & Hiebler, S. (2007). Modeling of subcooling and solidification of phase change materials. *Modell Simulat Mater Sci Eng*, (Dec 2007), Vol. 15, No. 7, pp. 879–892, 0965-0393.

Hamdan, M.A. & Elwerr, F.A. (1996). Thermal energy storage using a phase change material. *Sol Energy*, (Feb 1996), Vol. 56, No. 2, pp.183–189, 0038-092X.

Kürklü, A. (1998). Energy storage applications in greenhouses by means of phase change materials (PCMs): a review. *Renew Energy*, (Jan 1998), Vol. 13, No. 1, pp. 89-103, 0960-1481.

Lazaro, A.; Dolado, P.; Marín, J.M. & Zalba, B. (2009a). PCM-air heat exchangers for freecooling applications in buildings: experimental results of two real-scale prototypes. *Energy Conv Manag*, (Mar 2009), Vol. 50, pp. 439-443, 0196-8904.

Lazaro, A.; Dolado, P.; Marin, J.M. & Zalba, B. (2009b). PCM-air heat exchangers for freecooling applications in buildings: empirical model and application to design. *Energy Conv Manag*, (Mar 2009), Vol. 50, pp. 444-449, 0196-8904.

Lazaro, A. (2009). *Thermal energy storage with phase change materials. Building applications: materials characterization and experimental installation to test PCM to air heat exchanger prototypes*. Thesis, University of Zaragoza, Zaragoza, Spain.

Lazaro, A.; Zalba, B.; Bobi, M. & Castellón, C. (2006). Experimental Study on Phase Change Materials and Plastics Compatibility. *AIChE J*, (Feb 2006), Vol. 52, No. 2, pp. 804-808, 0001-1541.

London, A.L. & Seban, R.A. (1943). Rate of ice formation. *Transactions of the ASME*, Vol. 65, pp. 771–778.

Marin, J. M. & Monne, C. (1998). *Transferencia de calor (Heat transfer)*, Kronos, 8488502729, Zaragoza, Spain.

Marin, J.M.; Zalba, B.; Cabeza, L.F. & Mehling, H. (2003). Determination of enthalpy-temperature curves of phase change materials with the temperature-history method: improvement to temperature dependent properties. *Meas Sci Techno*, (Feb 2003), Vol. 14, No. 2, pp. 184-189, 0957-0233.

McKay, M.D.; Conover, W.J. & Beckman, R.J. (1979). A comparison of three methods for selecting values of input variables in the analysis of output from a computer code. *Technometrics*, Vol. 21, No. 2, pp. 239–245, 0040-1706.

Mehling, H. & Cabeza, L.F. (2008). *Heat and cold storage with PCM. An up to date introduction into basics and applications*, Springer-Verlag, 978-3-540- 68556-2, Berlin-Heidelberg, Germany.

Mills, A.; Farid, M.; Selman, J.R. & Al-Hallaj, S. (2006). Thermal conductivity enhancement of phase change materials using a graphite matrix. *Appl Therm Eng*, (Oct 2006), Vol. 26, No. 14-15, pp. 1652-1661, 1359-4311.

Pérez Vergara, I.G.; Díaz Batista J.A. & Díaz Mijares, E. (2001). Simulation experiments optimized by response surfaces, *Centro Azucar*, (Jan 2001), Vol. 2, pp. 68-74.

Sharma, A.; Tyagi, V.V.; Chen, C.R. & Buddhi, D. (2009). Review on thermal energy storage with phase change materials and applications. *Renew Sust Energ Rev*, (Feb 2009), Vol. 13, pp. 318-345, 1364-0321.

Turnpenny, J.R.; Etheridge, D.W. & Reay, D.A. (2001). Novel ventilation system for reducing air conditioning in buildings. Part II: testing of prototype. *Appl Therm Eng*, (Aug 2001), Vol. 21, No. 12, pp. 1203-1217, 1359-4311.

Watanabe, H. (2002). Further examination of the transient hot-wire method for the simultaneous measurement of thermal conductivity and thermal diffusivity. *Metrologia*, Vol. 39, No. 1, pp. 65-81, 0026-1394.

Yanbing, K.; Yi, J. & Yinping, Z. (2003). Modeling and experimental study on an innovative passive cooling system e NVP system. *Energy Build*, (May 2003), Vol. 35, No. 4, pp. 417-425, 0378-7788.

Zalba, B.; Marín, J.M.; Cabeza, L.F. & Mehling, H. (2004). Free-cooling of buildings with phase change materials. *Int J Refrig*, (Dec 2004), Vol. 27, No. 8, pp. 839-849, 0140-7007.

Zalba, B.; Marín, J.M.; Cabeza, L. & Mehling, H. (2003). Review on thermal energy storage with phase change: materials, heat transfer analysis and applications. *Appl Therm Eng*, (Feb 2003), Vol. 23, No. 3, pp. 251–283, 1359-4311.

Zhang, Y.; Jiang, Y. & Jiang, Y. (1999) A simple method, the T-history method, of determining the heat of fusion, specific heat and thermal conductivity of phase-change materials. *Meas Sci Techno*, (Mar 1999), Vol. 10, No. 3, pp. 201-205, 0957-0233.

Zhang, D.; Tian, S.L. & Xiao, D.Y. (2007). Experimental study on the phase change behavior of phase change material confined in pores. *Sol Energy*, Vol. 81, No. 5, pp. 653-660, 0038-092X.

Zukowski, M. (2007a). Experimental study of short term thermal energy storage unit based on enclosed phase change material in polyethylene film bag. *Energy Conv Manag*, (Jan 2007), Vol. 48, No. 1, pp. 166-173, 0196-8904.

Zukowski, M. (2007b). Mathematical modeling and numerical simulation of a short term thermal energy storage system using phase change material for heating applications. *Energy Conv Manag*, (Jan 2007), Vol. 48, No. 1, pp.155–65, 0196-8904.

The Soultz-sous-Forêts' Enhanced Geothermal System: A Granitic Basement Used as a Heat Exchanger to Produce Electricity

Béatrice A. Ledésert and Ronan L. Hébert
Géosciences et Environnement Cergy,
Université de Cergy-Pontoise
France

1. Introduction

The increasing need for energy, and electricity in particular, together with specific threats linked with the use of fossil fuels and nuclear power and the need to reduce CO_2 emissions leads us to look for new energy resources. Among them, geothermics proves to be efficient and clean in that it converts the energy of the earth into heating (domestic, industrial or agricultural purposes) or electricity (Lund, 2007). Numerous geothermal programs are producing energy at present and some of them have been performing for several decades in the USA (Sanyal and Enedy, 2011), Iceland and Italy for example (Minissale, 1991; Romagnoli et al., 2010). From statistics presented in World Geothermal Congress 2010, the installed capacity of geothermal power generation reaches 10,715 MW in the world. It increased by nearly 20% in 5 years. Its average annual growth rate is around 4%. USA, Indonesia and Iceland increased by 530MW, 400MW and 373MW respectively. Many countries all around the world develop geothermal exploitation programs. As a consequence, scientists from the whole world meet each year at the Annual Stanford Workshop on Geothermal Reservoir Engineering to discuss new advances in geothermics.

Conventional geothermal programs use naturally heated groundwater reservoirs. In many sedimentary provinces, depths of a few hundreds of meters are enough to provide waters with a temperature around 90°C. Such resources give rise to low and very low enthalpy geothermics. Very low enthalpy geothermal resources are used through geothermal heat pumps for various purposes including hot water supply, swimming pools, space heating and cooling either in private houses or in public buildings, companies, hotels and for snow-melting on roads in Japan (Yasukawa and Takasugi, 2003). In 1999 the energy extracted from the ground with heat pumps in Switzerland reached 434 GWh. The same level of utilization in Japan would bring the Japanese figure to 8 TWh per year (Fridleifsson, 2000). Technically, heat pumps can be applied everywhere. It is the difference between surface (atmospheric) and underground temperatures at 20 m or deeper that provides the advantage of geothermal heat pumps over air-source heat pumps.

In volcanic zones (like in Iceland), geothermics depends on specific geological contexts that are rather rare on the earth even though quite numerous in specific zones e.g. in the vicinity of subduction zones like around the Pacific Ocean as in Japan (Tamanyu et al., 1998) or in zones where the earth's crust is expanding like in Iceland (Cott et al., 2011). Such geothermics is called high enthalpy. It allows the production of electricity like in the Uenotai geothermal power plant in Japan which started operation in 1994 as a 27.5 MW electric power generation facility (Tamanyu et al., 1998). Electricity production from geothermal resources began in 1904 in Italy, at Larderello (Lund, 2004; Massachusetts Institute of Technology [MIT], 2006). Since that time, other hydrothermal developments led to an installed world electrical generating capacity of nearly 10,000 MWe and a direct-use, nonelectric capacity of more than 100,000 MWt (thermal megawatts of power) at the beginning of the 21st century from the steam field at The Geysers (California, USA), the hot-water systems at Wairakei (New Zealand), Cerro Prieto (Mexico), Reykjavik (Iceland), Indonesia and the Philippines (MIT, 2006).

Complementary to conventional geothermics, Enhanced Geothermal Systems (EGS; also called Engineered Geothermal Systems) aim to develop reservoirs in rocks where little (or no) water is available (Redden et al., 2010). This concept was invented, patented and developed in the early 1970s at Los Alamos National Laboratory and was first called Hot Dry Rock (HDR) geothermal energy. As defined by these early researchers, the practical HDR resource is the heat contained in those vast regions of the earth's crust that contain no fluids in place – the situation characterizing by far the largest part of the earth's drilling-accessible geothermal resource (Brown, 2009).

This concept was developed for electricity production in any kind of area at the surface of the earth even though the geodynamical context is not in favour of geothermics (Redden et al., 2010 and references therein).

However, because of the general low thermal gradient in the earth (30°C/km in sedimentary basins), reaching a temperature around 150-200°C needed for the production of electricity make things more difficult than first considered. Technical and economical problems linked to such deep drillings (Culver, 1998; Rafferty, 1998) have restricted EGS to zones where the thermal gradient is high enough to reduce the depth of the exchanger. Now, EGS include all geothermal resources that are currently not in commercial production and require stimulation or enhancement (MIT, 2006).

Table 1 gives an overview of HDR/EGS programs in the world.

In such difficult technical conditions, one can wonder whether the geothermal energy resource and electricity production process are sustainable. According to Clarke (2009) and authors cited therein, the management and use of the geothermal resource (Rybach and Mongillo, 2006) and the environmental impacts during geothermal energy production (Bloomfield et al., 2003; Reed and Renner, 1995) were the first concern about sustainability of geothermal energy. These studies have shown that there is less impact on land use, air emissions including greenhouse gases, and water consumption from geothermal electricity generation than from fossil-fuel–based electricity generators. However, the environmental impacts from the construction of geothermal energy production facilities being less well understood, especially for enhanced geothermal systems (EGS) subsequent studies were conducted. The life-cycle analysis of the EGS technology (including pre-production process such as drilling, construction, production and transportation) had to be discussed, especially when the potential for large-scale development exists. Because of increased depth and decreased water availability, environmental impacts may be different from those of

Country	Site	Depth (m)	Dates	Production of electricity
France	Le Mayet		1975-1989	
	Soultz-sous-Forêts	5000 (3 boreholes)	1985-	1.5 MWe Since June 2008
Germany	Bad Urach	4500 and 2600	1976-	
	Falkenberg	300	1975-1985	
UK	Rosemanowes	2700	1975-1991	
Australia	Cooper Basin	4300	2001-	
USA	Fenton Hill	5000	1973-2000	
	Coso		2001-	
	Desert Peak	5420	2001-	
Japan	Hijori	2300	1987-	

MWe: MWelectricity (raw production minus consumption of electricty required for the production).
Data after MIT (2006), Davatzes and Hickmann (2009) and Genter et al. (2009)

Table 1. Overview of some HDR or EGS programs in the world.

conventional geothermal power generation. It is expected that EGS will produce less dissolved gas (mostly carbon dioxide CO_2 and hydrogen sulfide H_2S) than conventional energy by recovering the heat through a heat exchanger and reinjecting the fluid without releasing any gas during operation. As concerns subsurface water contamination it is unlikely because all the produced fluid is reinjected. EGS is also characterized by a modest use of land since with directional-drilling techniques, multiple wells can be drilled from a single pad to minimize the total wellhead area (MIT, 2006). EGS requires no storage and the plant is built near the geothermal reservoir because long transmission lines degrade the pressure and temperature of the geofluid as well as the environment. As a consequence EGS power plants require about 200 m²/GWh while a nuclear plant needs 1200 m²/GWh and a solar photovoltaic plant 7500 m²/GWh (MIT, 2006). Most of EGS developments are likely to occur in granitic-type crystalline rocks, at great depth. Careful management of the water resource is unlikely to induce subsidence (lowering of the ground's level as in shallow mining activity; MIT, 2006). Seismic activity linked to engineering of EGS reservoirs during hydraulic stimulation (injection of water under pressure to create or open pre-existing joints) has conducted managers of these sites to prefer chemical stimulations (use of chemicals to dissolve minerals responsible for the sealing of joints) in or close to urban areas (Hébert et al., 2011; Ledésert et al., 2009) in order to avoid earthquakes. Another force of the EGS technology is the possible use of CO_2 instead of water because of its favourable thermodynamic properties over water in EGS applications (Brown, 2000; Magliocco et al., 2011) thus leading to a possible sequestration of carbon dioxide produced by the use of fossil fuel. However EGS might have increased visual impact and noise levels compared to conventional geothermal power plants but no more than fossil fuel driven power plants (MIT, 2006). The highest noise levels are usually produced during the well drilling, stimulation, and testing phases (about 80 to 115 decibels). For comparison, congested urban areas typically have noise levels of about 70 to 85 decibels, and noise levels next to a major freeway are around 90 decibels. A jet plane just after takeoff produces noise levels of about 120 to 130 decibels (MIT, 2006). Finally, considering all these factors, EGS has a low overall environmental impact when the production of electricity is considered compared to fossil or nuclear generation (MIT, 2006). As concerns the demand, supply and economic point of view, MIT (2006) provides a rather detailed analysis.

Geothermal energy and EGS in particular is studied world-wide and the annual Workshop on Geothermal Reservoir Engineering (Stanford, California, USA) allows scientists and industrials to compare data and improve renewable energy production. Proceedings of the workshop can be downloaded very easily and for free on the Internet. As a consequence, the latest advances in geothermal technology are available to the scientific community.

2. EGS technology

Flow rates on the order of 50 L.s^{-1} and temperatures of 150° C to 200° C are required to allow an economical generation of electrical energy from geothermal resources (Clauser, 2006). Heat source risk can be quantified via a detailed assessment of surface heat flow together with measurements of temperature in boreholes (for example, for the Desert Peak Geothermal area, at 0.9 to 1.1 km depth, the ambient temperatures is of ~180 to 195° C in rhyolite tuff and argillite; Hickman and Davatzes, 2010). Thermal insulation (measured on core or cuttings samples) together with precision surface heat flow measurements allows the prediction of temperature distribution at depth in one, two or three dimensions (Beardsmore and Cooper, 2009). For regions outside natural steam systems and high surface heat flow (for example Iceland, Indonesia, Turkey, etc.) conditions necessary for electricity production are met at depths below 3 km provided the underground rock heat exchanger is engineered in order to increase the paths available for the fluid flow. Such systems are called Engineered (or Enhanced) Geothermal Systems (EGS) or Hot Dry Rock (HDR). In these techniques, the host rock is submitted to stimulations (Economides and Nolte, 2000) in order to increase the heat exchange surface between the rock and the injected fluid. Stimulations are derived from petroleum technology where they have been used for decades. The first method consists in the injection of water under high pressure to create irreversible shearing and opening of fractures and is called hydraulic stimulation (Portier et al., 2009). The second method, called chemical stimulation (e.g. Nami et al., 2008; Portier et al., 2009) uses various kinds of chemical reactants to dissolve minerals and to increase permeability. Both methods have proven successful for enhancing permeability at depth but it is still a challenge to plan and control the stimulation process. Details about stimulations can be found in the abundant litterature (e.g. Kosack et al., 2011; Nami et al., 2008; Portier et al., 2009 and references therein). During or after stimulations, tracers are used to assess the connectivity between the wells and the speed of fluid transfer. Many examples of use of tracers are found in the literature (e.g. Radilla et al., 2010; Redden et al., 2010; Sanjuan, 2006). In addition, prior to any stimulation or circulation test between the wells, in-situ stress and fracture characterization have to be considered with great attention in order to better constrain the geometry and relative permeability of natural or artificially created fractures (e.g. Hickman and Davatzes, 2010 for the Desert Peak geothermal field). A subsequent modelling of the 3D fracture network (Genter et al., 2009; Sausse et al., 2010, Dezayes et al., 2011) and of flow and transport along the fractures can be profitably performed (e.g. Karvounis and Jenny, 2011) to predict the behaviour of the thermal exchanger and ensure its financial viability. Such modelling is based on the accurate knowledge of the fracture network obtained through seismic records performed during stimulation or production tests (Concha et al., 2010) and thanks to thorough characterization of the fracture network (Hébert et al., 2010, 2011; Ledésert et al., 2009). When the rock heat exchanger is finally operated, careful reservoir engineering and monitoring has to be performed to ensure the viability of the EGS (Satman, 2011). The produced hot fluid is continuously replaced by cooled injected water. After the thermal breakthrough time the temperature of the produced fluid decreases. However if after a time the field is shut-in the natural energy flow will slowly replenish the geothermal system and it will again be available for production. Therefore when operated on a periodic basis, with production followed by recovery, doublets are renewable and sustainable (Satman, 2011). Triplets (one injection and two production boreholes) are now considered as being the best configuration (MIT, 2006; Genter et al., 2009). However, it must be taken into consideration that when an EGS reservoir is developed through hydraulic fracturing, the size of the reservoir might extend too much and attendant high water losses might occur compromising the sustainability of the project as at Fenton Hill (MIT, 2006; Brown, 2009). When shearing occurs through reopening of pre-existing sealed fractures during hydraulic stimulation (e.g. at Fenton Hill; Brown, 2009) or when the mineral deposits are dissolved through chemical stimulation (e.g. at Soultz; Nami et al., 2008; Portier et al., 2009; Genter et al., 2009) the size of the exchanger is better constrained and fluid losses are limited.

Other risks such as technology (reliable supply of produced geofluids with adequate flow rates and heat content), finances (cost of construction, drilling, delays), scheduling, politics, etc…have to be estimated in order to make an EGS project viable. They are presented in MIT (2006) for the different stages of a project. In addition, seismic risk has to be fully taken into account where EGS programs are to be developed in urban areas in order to produce both electricity and central heating (Giardini, 2009): the Basel (Switzerland) experience (see section 4.4) had to be stopped because of a 3.4 magnitude earthquake generated by stimulation in a naturally seismic area. As a consequence to these numerous constraints, no financially viable EGS program is operating at present but the production of electricity that began at Soultz-sous-Forêts (France) in June 2008 is highly promising.

As a conclusion, Table 2 shows some forces and difficulties of EGS programs inferred from the literature.

Forces	Difficulties
Production of electricity	Deep drilling (technical and financial difficulties)
Sustainability of the resource provided its correct management	Risk of water loss in case of pure hydraulic fracturing
Low to no GHG emissions	Engineering of the reservoir to increase permeability
Low global environmental impact compared to fossil-fuel and nuclear electricity	Adequate flow rate and temperature
Available on continents worldwide	No financially viable program operating at present

Table 2. Forces and difficulties of EGS programs inferred form the abundant literature on the subject (see reference list). GHG: greenhouse gases.

3. The Soultz-sous-Forêts EGS

The Soultz-sous-Forêts' (called Soultz in the following) project began in the late eighties thanks to a particular geological context. Several zones in France are submitted to high heat flows, among which the Soultz area, because of the development of a rift system in northern Europe (Figure 1). Initiated by a French-German team (Gérard et Kappelmeyer, 1987), the Soultz program has been a European project with a significant Swiss contribution mainly supported by public funding between 1987 and 1995 and co-funded by industry from 1996 to present (Genter et al., 2009). The Soultz projetc represents a multinational approach to develop an EGS in Europe.

3.1 Geological context

The Soultz EGS site is located in the upper Rhine graben (Figure 2) where a high heat flow was measured at shallow depth in old oil wells (110 °C/km). Natural water being found in great amount at depth in the granitic heat exchanger, the project was not HDR anymore and was then called EGS since numerous stimulations (first hydraulic and then chemical with several steps and chemical reactants) were necessary to improve the connection between the 3 deep wells (around 5 000 m deep). First investigations of the fracture network showed that they are grouped in clusters separated by little or no-fractured zone, following a fractal organisation (Ledésert et al., 1993).

Figure 2 shows the more detailed location of the Soultz site and extension of the thermal anomaly related to the upper Rhine graben(URG) also showed on a geological cross section on which the Soultz horst can be distinguished.

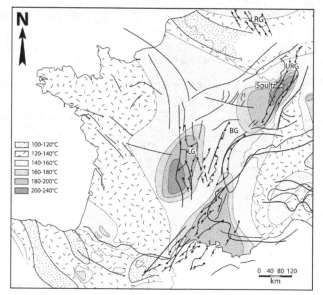

Fig. 1. Map of extrapolated temperatures at 5 km depth and location of some major structural structures (modified after Hurtig et al., 1992 and Dèzes et al., 2004). LRG : Lower Rhine Graben; URG : Upper Rhine Graben; BG : Bresse Graben; LG : Limagne Graben.

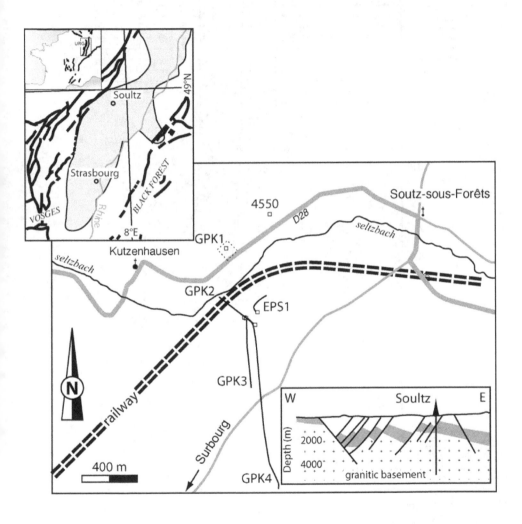

Fig. 2. Location map of the upper Rhine graben (URG) in eastern France and of the Soultz-sous-Forêts' site, 50 km North of Strasbourg, in a zone of high thermal anomaly (grey on the map). Six boreholes are present : 4550 (previous oil well), GPK1 (first HDR borehole), EPS1 (entirely cored scientific HDR borehole), GP2-GPK3-GPK4 (5 000m deep boreholes forming the triplet of the EGS). Their horizontal trajectories are shown on the main map. The E-W geological cross section shows the geometry of the upper Rhine graben and of the Soultz horst. Dots represent the granite while inclined grey and white layers correspond to the sedimentary cover. After Ledésert et al. (2010).

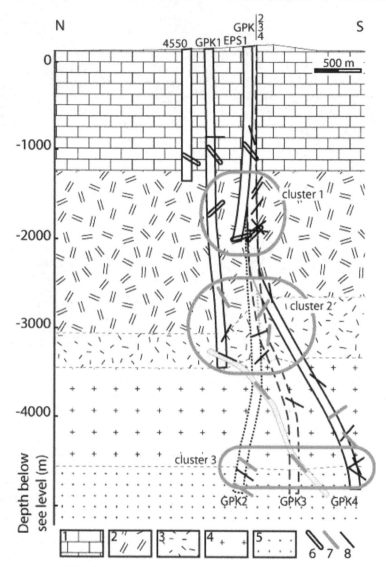

Fig. 3. Cross-section of the Soultz geothermal system. Note 3 zones intensely fractured and altered by natural fluids, noted cluster. A major drain is encountered in GPK1 near 3200m and is found in GPK2 around 3500m and GPK3 close to 4500m (represented with a light-grey curvated wide line). 4550 (oil drill hole); EPS1 (cored scientific hole); GPK1 (scientific hole, destructive conditions, few core pieces). 1: sedimentary cover, 2: standard porphyritic Bt-Hbl granite, 3: standard granite with fractures and vein alteration, 4: Bt+Hbl - rich granite becoming standard granite at depth, 5: two-mica and Bt-rich granite, 6: Level 1 fracture, 7: Level 2 fracture, 8: Level 3 fracture. Figure modified after Dezayes and Genter (2008) and Hébert et al. (2011). Mineral abbreviations according to Kretz (1985), Bt : biotite, Hbl : hornblende.

3.2 Development of the Soultz EGS

The granite body is used as a heat exchanger in which the fracture planes are surfaces of heat exchange between the injected water and the hot rock mass. As a consequence, the knowledge about the fractures is necessary to better understand and predict the behaviour of the heat exchanger. This is why numerous studies have been performed on the natural fracture network (Ledésert et al., 1993; Genter et al., 1995; Sausse et al., 2007, 2008, 2009; Dezayes et al., 2008; Dezayes et al., 2011). Dezayes et al. (2010) have classified the fracture zones into three different categories (or levels) on the basis of their relative scale and importance as fluid flow paths (see complete review in Dezayes et al., 2010). Level 1 corresponds to major fracture zones, which were permeable prior to any stimulation operation (Figure 3) and were subject to important mud loss during the drilling operation. Fracture zones of level 2 are characterized by at least one thick fracture with a significant hydrothermal alteration halo. They showed a flow indication higher than 20% of fluid loss during stimulation. Fracture zones of level 3 show a poorly developed alteration halo and a fluid loss below 20% during stimulation. Figure 3 shows the location of the boreholes and that of fracture clusters that were reactivated during stimulations allowing connection between the wells.

According to the abundant literature on the subject, developing an EGS is a difficult and costly task that deserves thorough studies. The knowledge of the geometry of the fracture network being crucial as explained before, many methods are used to improve it. Concha et al. (2010) indicate that microseismic data can be used profitably in that they provide information about reservoir structure within the reservoir rock mass at locations away from boreholes, where few methods can provide information. They used microseismic events induced from production and hydrofracturing tests performed in 1993 as sources for imaging the Soultz EGS. These tests injected 45,000 m^3 of water at depths between 2850 and 3490 m and resulted in over 12,000 microseismic events that were well recorded by a four station downhole seismic network. Concha et al. (2010) began by determining a three dimensional velocity model for the reservoir using Double Difference Tomography for both P and S waves. Then they analyzed waveform characteristics to provide more information about the location of fractures within the reservoir. Using such methods, it appears that the volume of the exchanger stimulated during operation of the Soultz EGS is approximately $1km^3$.

The Soultz EGS is characterized by three deep boreholes (CPK2, GPK3 and GPK4; ca 5000 m; Figure 3). They were drilled after GPK1 and EPS1 boreholes that could not be used for the EGS development because of technical problems, and oil wells such as 4550 (Figures 2 and 3).

Genter et al. (2009) provide an overview of the Soultz project. The first exploration of the geothermal Soultz site consisted in exploration by drilling at shallow depth (GPK1, 2 km). Then convincing results were obtained between 1991 and 1997 through a 4 month circulation test successfully achieved between 2 wells in the upper fractured granite reservoir at 3.5 km. Based on these encouraging results, 3 deviated wells (GPK2, GPK3, GPK4) were drilled down to 5 km depth between 1999 and 2004 for reaching down-hole temperatures of 200°C (Genter et al., 2009). They form the geothermal triplet. Geothermal water is pumped from the production wells (GPK2, GPK4) and re-injected together with fresh surface water at lower temperature into the injection well GPK3. On a horizontal view, the 3 deep deviated wells are roughly aligned along a N170°E orientation (Figure 2) corresponding to the orientation of both the main fracture network and present-day principal maximal horizontal stress, allowing the best recovery of the injected water. The three deep boreholes were drilled from the same platform, about 6 m apart at the surface whereas at their bottom, the distance between each production well and the re-injection well is about 700 m. The 3 wells are cased between the surface and about 4.5 km depth offering an open-hole section of about 500 m length (Genter et al., 2009).

	GPK3 (injection well)	GPK2	GPK4
Pumping rates (L/s)	15	11.9	3.1
Arrival time for fluorescein (days)	injection	4	24
Volume of fluid	209 000	165 000	40 000
Permeability relative to GPK3 (m^2)		10^{-13}	10^{-15}
Quality of connection with GPK3		High	Low

Table 3. Results of circulation tests between the three deep wells showing the strong discrepancy between the two production wells, GPK2 and GPK4. Data in Sanjuan et al. (2006); Genter et al. (2009) and Kosack et al. (2011).

The geothermal wells were stimulated (hydraulically and chemically) between 2000 and 2007 in order to enhance the permeability of the reservoir that was initially low (Table 3) in spite of a large amount of fractures (up to 30 fractures/m; Ledésert et al., 1993; Genter et al., 1995). Figure 4 provides a synthetic view of the increase in the productivity/injectivity rates for each of the Soultz deep boreholes after hydraulic and chemical stimulations. A 5-month circulation test, carried out in 2005 in the triplet, showed similar results as in 1997 in terms of hydraulics (Nami et al., 2008): in both cases, a recovery of about 30% of the fluid mass was obtained at the production wells showing the open nature of the reservoir (Gérard et al., 2006). This result is opposed to the HDR concept where the reservoir is closed (Brown, 2009) and no water naturally exists in the reservoir prior to its injection. The limited recovered mass of injected fluid was continuously compensated by native brine indicating direct connections with a deep geothermal reservoir (Sanjuan et al., 2006). To give an example of stimulation test, from July to December 2005, about 209 000 m^3 of fluid were injected into GPK3 and 165 000 m^3 and 40 000 m^3 were produced from GPK2 and GPK4 respectively (Sanjuan et al., 2006), yielding a nearly even mass balance. In addition, a mass of 150 kg of 85 % pure fluorescein was dissolved in 0.95 m^3 of fresh water and was used as a tracer injected into GPK3 over 24 hours, while geochemical fluid monitoring started at GPK2 and GPK4. Fluorescein was first detected in GPK2, 4 days after the injection into GPK3. In GPK4, fluorescein was detected only 24 days after the injection. The average pumping rates were 11.9 L.s^{-1} in GPK2, 15 L.s^{-1} in GPK3, and 3.1 L.s^{-1} in GPK4, already indicating a reduced water supply to GPK4 (Sanjuan et al., 2006; Genter et al., 2009). These results show that the hydraulic connection is very heterogeneous: it is rather easy between GPK3 and GPK2 while it is much more difficult between GPK3 and GPK4 (Table 3). The permeability in most of the reservoir is on the order of 10^{-17} m². A good connection is naturally established between GPK2 and GPK3 with a mean permeability on the order of 10^{-13} m², while a barrier exists to GPK4 (Kosack et al., 2011).

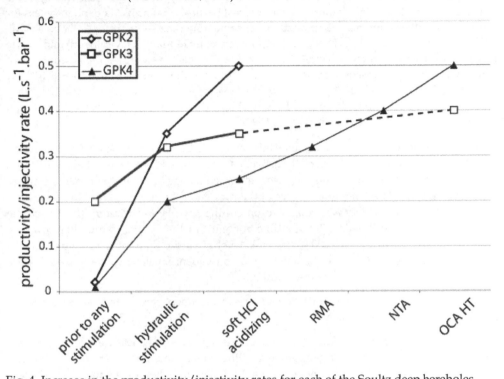

Fig. 4. Increase in the productivity/injectivity rates for each of the Soultz deep boreholes after hydraulic and various chemical stimulations (after Hébert et al., 2011). GPK3 had an initial rate higher than GPK2 and GPK4 (prior to any stimulation data). Because of its very low initial rate and different behaviour, GPK4 had to face multiple chemical stimulations and finally reached the same level as GPK2 even though its rate was only half that of GPK2 after the soft HCL acidizing. RMA : regular mud acid; NTA : nitrilo-triacetic acid (chelating agent); OCA HT : organic clay acid for high temperature.

Following processes run in oil-production wells to improve the permeability of a rock reservoir, two basic types of chemical stimulations can be conducted: matrix acidizing and fracture acidizing. Matrix stimulation is accomplished, for example in sandstones, by injecting a fluid (e.g. acid or solvent) to get rid of materials that reduce well productivity or injectivity. Fracture acidizing is used to develop conductive paths deeper into the formation. This treatment consists of injecting an acid fluid into the formation at a rate higher than the reservoir matrix will accept. This rapid injection produces a wellbore pressure build-up leading to a fracturing of the rock. Continued fluid injection increases the fracture's length and width (Portier et al., 2009).

At Soultz, thorough petrographic studies of the fracture network (Dubois et al., 2000; Genter et al., 1995; Hébert et al., 2010, 2011; Ledésert et al., 1999, 2009, 2010) have shown that more than 90% of the fractures are sealed by minerals that precipitated because of natural fluid flow. K-Ar dating of illite (K-bearing clay mineral) found in a fractured and altered zone located at 2200 m in the Soultz granite (Bartier et al., 2008) indicate at least two episodic illitization at 63 Ma or slightly more for the coarsest particles and at 18 Ma or slightly less for the smallest. Other minerals precipitated at the same time (Bartier et al., 2008; Dubois et al., 2000; Ledésert et al., 1999), such as tosudite (Li-bearing mixed-layer clay mineral) and calcite (calcium carbonate). Calcite precipitated from the Ca-ions liberated by the dissolution of primary plagioclases present in rather great abundance in the granite (nearly 40% of 10%-Ca oligoclase; Ledésert et al., 1999; Table 4) during hydrothermal flow and from sedimentary brines enriched in Ca-ions during its flow within the calcareous Muschelkalk layers that penetrated into the granite (Ledésert et al., 1999).

Recently we have focused on calcite since this mineral is thought to impair the permeability between the 3 deep wells and especially between GPK3 and GPK4 (Hébert et al., 2010, 2011; Ledésert et al., 2009, 2010). To this aim, we performed a compared study of calcite-content, other petrographic data (alteration degree of the rock and illite content) and fluid flow from well-tests. Petrographic data were obtained on cuttings by mano-calcimetry (for the calcite content) and X-Ray diffraction (for the illite content). In the Soultz granite, like in many other granites (Ledésert et al., 2009), the base level of calcite amount is around 1.8 wt % (Hébert et al., 2010; Ledésert et al., 2009). As a consequence, calcite contents over 2% are considered as calcite anomalies by these authors.

The three deep wells show distinct behaviours in the deep part of the exchanger (open holes, below 4500m depth; Table 5).

In GPK2, two main groups of fracture zones are distinguished. The less conductive ones are characterized by low alteration facies, moderate illite content and low calcite content (below 2 wt.%) likely resulting from the early pervasive fluid alteration. It suggests that these fracture zones are poorly hydraulically connected to the fracture network of the geothermal reservoir. On the opposite, the fracture zones with the best conductivities match with high to moderate calcite anomalies (respectively 11, 8, ~5 wt.%), high to moderate alteration grade and high illite content. This suggests massive precipitation of calcite from later fluid circulations within the fractured zone. Thus, the calcite content seems possibly proportional to conductivity.

In GPK3, the less conductive fracture zones are concentrated in a zone that extends from ~ 4875 to ~ 5000 m measured depth (MD), where they correlate with a large and high calcite anomaly zone. The main fracture zone, which accommodates 63–78% of the fluid flow, has the lowest calcite anomaly (2.9 c wt.%) of all the fracture zones of this well. Nearly all the moderate calcite anomalies occur in the vicinity of fracture zones. In this well, regarding the fracture zones data and the calcite anomalies, it seems that the more calcite the less fluid flow and therefore calcite plays a major role in the reduction of the conductivity of the fracture zones of this well. Thus, in GPK3, the maximum fluid flow and significant calcite deposit are not correlated as it is observed in the open-hole section of GPK2.

The highest calcite anomaly of all three deep wells is found in GPK4 (18%). In GPK4, the fluid flow is mainly accommodated via a single zone. All the other fracture zones are considered to have a similarly low fluid flow and are characterized by moderate or high

Site	Habanero (Cooper Basin, Australia)	Hijori (Hijori caldera, Japan)	Soultz (Rhine graben, France)
Rock type	Granite	Tonalite/granodiorite	Granite
Quartz	39.3	35.3	28.4
Plasioclase	29.7	38.2	39.9 (oligoclase)
K-feldspar	18.1	2.1	18.8
Muscovite/Biotite	8.4	0.4	8.4 (biotite)
Carbonate	1.1	1.3	*≤1.8-18*
Chlorite/Clay M	0	6.5	*<1*
Sericite	0	9.5	*up to several % (illite)*
Pyroxene	2.2	0	4.5 (amphibole)
Epidote	0	1.8	*< 1*
Calcopyrite	1.1	0	0
Anhydrite	0	4.9	0
Total	100	100	100, depending on the zones

Table 4. Mineral composition of EGS rock bodies (mostly expressed in volume %). Data
from Ledésert et al. (1999) and Yanagisawa et al. (2011). At Soultz, some zones are strongly
fractured and altered by natural hydrothermal fluids. In such zones, the composition of the
granite is strongly modified: primary quartz has been totally dissolved, oligoclase is
replaced by illite or tosudite (clay minerals), biotite and amphibole by chlorite and epidote.
Newly-formed minerals are indicated in italics.

	GPK2	GPK3	GPK4
Highly conductive fractures	High alteration High illite High calcite	Low calcite	Low calcite
relationship	Calcite proportional to conductivity : calcite is found in highly conductive fractures	The less calcite, the more fluid flow : calcite reduces conductivity	The less calcite, the more fluid flow : calcite reduces conductivity
Permeability	high	high	low
Connectivity	high	high	low

Table 5. Relationships between the amount of calcite and the intensity of fluid flows in the
three deep Soultz wells. Comparison with permeability data (from table 3). The connectivity
between the wells is deduced. No petrographic data (alteration degree and illite content) are
available for GPK3 and GPK4 because of poor quality cuttings.

calcite anomalies. Therefore it seems that in GPK4, the highest the fluid flow, the lowest the
calcite anomaly, as in GPK3.

However, GPK3 shows a high permeability while that of GPK4 is low (Table 5). Combining data of calcite content and permeability, one can infer that calcite may represent a serious threat to the EGS reservoir when the connectivity of the fractures is low while it does not impair the permeability when the connectivity is high. A solution can be brought by hydraulic fracturing that allows developing the extension of fractures. However, such process was employed in Basel (Switzerland) resulting in an earthquake of a 3.4 magnitude that scared the population in 2006. The EGS Basel project had to be stopped. At Soultz, an earthquake of 2.9 magnitude had been felt by local population during the stimulation of GPK3 in 2000 thus no further hydraulic stimulations were driven to prevent this problem. As a consequence, chemical stimulations had to be performed in order to improve the permeability and connectivity of the three deep wells. Particular efforts were put on GPK4. Figure 4 shows the results of chemical stimulations. The behaviour of the 3 deep wells has been largely improved. Given the good results of the circulation test conducted in 2005, and the improvement of the hydraulic performances of the three existing deep wells by stimulation, it was decided to build a geothermal power plant of Organic Rankine Cycle (ORC) type (using an organic working fluid). Thus, a first 1.5 MWe (electricity; equals 12 MW thermal) ORC unit was built and power production was achieved in June 2008 thanks to down-hole production pumps. The power plant was ordered to a European consortium made of Cryostar (France) and Turboden, Italy. A three year scientific and technical monitoring of the power plant has started on January 2009 focused on the reservoir evolution and on the technologies used (pumps, exchanger; Genter et al., 2009).

3.3 Technical data about the heat exchanger and the EGS (after Genter et al., 2009 and Genter et al., 2010)

The geothermal fluid is produced from GPK2 and GPK4 thanks to two different kinds of pumps and, after electricity production (or only cooling if electricity is not produced), it is reinjected in the rock reservoir through GPK3 and GPK1.

3.3.1 Pumps

It was necessary to install down-hole production pumps because the artesian production was not sufficient. Thus, two types of production pumps were deployed in the production wells: a Line Shaft Pump (LSP; in GPK2) and a Electro-Submersible Pump (ESP; in GPK4).

The LSP itself is in the well while the motor is at surface. The connection is obtained through a line shaft. The main advantage is to avoid installing the motor in hot brine, but the possible installation depth is limited and the line shaft has to be perfectly aligned. The LSP was supplied by Icelandic Geothermal Engineering Ltd. The length of the shaft is 345 m. The shaft (40 mm diameter) is put in an enclosing tube (3" internal diameter) with bearings every 1.5 m. The enclosing tube is set by means of centralisers in the middle of the LSP production column (6" internal diameter) which is put into the 8" casing. The pump itself is from Floway (USA) and made of 17 different stages of 20 cm (3.4 m total length). The LSP flow rate can be modulated until 40l/s with a Variable Speed Drive. The maximum rotation speed is 3000 rpm at 50 Hz. The surface motor is vertical. Metallurgy is cast iron and injection of corrosion inhibitor can be done at the pump intake by mean of coiled tubing. Shaft lubrication is made with fresh water injected from surface in the enclosing tube. The pump has been installed at 350 m depth into GPK2 that presents good verticality and is the best producer. Due to hydraulic drawdown, the maximum flow rate expected with the LSP installed at 350 m is 35 l/s. During summer 2008, (07th July to 17th August), after six weeks of geothermal production (25 l/s, 155°C), scaling problems were observed within the lubrication part of the shaft. The fresh water used for lubricating the shaft was too mineralized and some carbonate deposits (calcite, aragonite) precipitated. Then, a poor lubrication occurred and the first axis of the shaft broke. Between mid August and November 2008, both the shaft and the pump were fully dismantled, analyzed and a demineralization water system was set up. The LSP pump was re-installed at 250 m depth in GPK2 and worked properly afterwards.

Both the ESP pump and its motor are installed into the GPK4 well at 500 m depth. The maximum expected flow rate from GPK4 equipped with ESP is 25 l/s but the pump is designed to a maximum flow rate of 40 l/s. The ESP was delivered by Reda/Schlumberger. Due to the expected maximum temperature (185°C) and the salty composition of the brine, specific design and noble metallurgy had to be used. The electrical motor is beneath the pump and connected to it through a seal section that compensates oil expansion and metallic dilatation. The motor is cooled by the pumped geothermal brine and internal oil temperature can reach 260°C. A fiber optic cable has been deployed with the ESP and allows monitoring the motor temperature and gives downhole information about the geothermal draw-down in the well. The first production tests from GPK4 with the ESP with an expected target of 25 l/s started on mid November 2008. After some days of production, GPK4 production decreased to 12.5 l/s at 152°C and the geothermal water was re-injected in GPK3 at 50°C. GPK2 flow rate was stabilized at 17.5 l/s for a temperature around 158°C. Both flows coming from GPK2 and GPK4 were re-injected under full automatism in GPK3 at 30 l/s. The ORC commissioning started for these geothermal conditions at around 155°C. GPK3 well-head pressure was maintained around 70-80 bars for reinjection.

3.3.2 Heat exchanger

A schematic view of the Soultz' binary power plant is given in Figure 5. As the purpose of the project was first to demonstrate the feasibility of power production, a binary system utilizing an organic working fluid called an Organic Rankine Cycle (ORC) technology was chosen. Due to the high salinity of the geothermal brine, the geothermal fluid cannot be vaporized directly into the turbine as occurs in classical "simple flash" power plants.

Then, a secondary circuit is used that involves a low boiling point organic working fluid (isobutane). As there is no easily accessible shallow aquifer around the geothermal site, an air-cooling system was required for the power plant, which also limits the impact on environment. It consists in a 9-fan system. The turbine is radial and operates around 13000 rpm. The generator is asynchronous and is running around 1500 rpm. The generator is able to deliver 11 kV and the produced power is to be injected into the 20 kV local power network.

The expected net efficiency of the ORC unit is 11.4%. Geothermal water may be cooled down to 80-90°C in the heat exchangers of the binary unit. After this cooling, the entire geothermal water flow rate is re-circulated in the reservoir. The system is built so that the production coming from one or two wells can easily be used to feed the power production loop. On surface, the pressure in the geothermal loop is maintained at 20 bars in order to

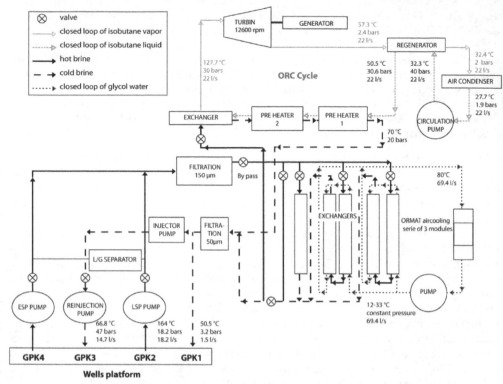

Fig. 5. Schematic view of the Soultz' binary power plant after Genter et al. (2010). Each production well can be run separately thanks to the valves. The hot geothermal fluid (around 165°C) is filtered (150 μm) before entering the surface network. After the complete cycle, the fluid is reinjected in the natural rock exchanger thanks to one or two wells : with a reinjection pump in GPK3 and in addition if necessary by gravity in GPK1. The temperature, pressure and flow figures indicated are those obtained during the 8 months circulation test performed in 2008. If the Organic Rankine Cycle dedicated to electricity production is not activated (the last valve being closed), the geothermal fluid goes through 5 exchangers in the cooling cycle (lower part of the figure with the ORMAT aircooling system) in order to be reinjected after filtration (50 μm) at low temperature (around 50 to 67°C). The last of these five exchangers is used only when necessary.

avoid mineral precipitations. Locally, in the filtering system, some scaling was observed with barite, celestine, iron oxides, galena and calcite mainly. In order to investigate corrosion and scaling, an innovative corrosion pilot was set up on the surface geothermal loop and tested for the first time between September 2008 and February 2009. Different kinds of steel were investigated for corrosion in the geothermal conditions of re-injection (20bars, < 80°C).

The liquid hot brine is pumped from the rock reservoir and first filtered in the surface geothermal loop in a self-cleaning 150μm filter. Whether the ORC cycle is or not in function, either the geothermal fluid feeds the ORC exchanger to produce electricity or it feeds the

Barriquand's exchangers to cool the brine to be reinjected. In that last case, the brine is injected in the 5 exchangers (Fig. 5). The fifth exchanger is used only when needed. The exchangers allow heat transfer between the brine and a fluid composed of water and mono propylene glycol.

3.4 Review of 20 years of research at Soultz

About 40 PhD theses have been written on the Soultz project in the last 20 years together with about 200 publications in international journals between 2001 and 2008. With this scientific background and the current production of electricity (1.5 MWe), the Soultz site is now a world reference for EGS. It appears that cooling 1 km^3 of rock by only 20°C (initial temperature around 160-200°C) liberates as much energy as the combustion of 1 275 000 tons of oil and thus saves as much non-renewable fossil fuel (SoultzNet, 2011).

3.5 Future of the Soultz EGS

The total cost of the Soultz pilot operating now is 54 M€ (Soultznet, 2011). A prototype of 20-30 MWe will follow the first pilot presently in production (1.5 MWe). On a longer term, industrial units will be constructed (Soultznet, 2011). Large-scale production units inspired from the Soultz EGS might transform the world of energy since it is clean and sustainable. It preserves fossil fuels and limits the emissions of GHG and allows a continuous production of electricity 8000 hours/year, at night as well as at day, whatever the climate conditions (SoultzNet, 2011).

4. Other EGS programs in the world

The Soultz EGS is the only operating site at present. It benefited from the experience developed on other sites all around the world. The objective here is not to provide a complete review of these projects but to show the impact they had on the Soultz project. More details on these projects can be found in MIT (2006) and in the abundant literature easily available (e.g. Brown, 2000, 2009; Yanagisawa et al., 2011; Yasukawa and Takasugi, 2003). An overview of HDR/EGS programs in the world is given in table 6.

4.1 Fenton Hill (U.S.A.; after Brown, 2000, 2009; MIT, 2006)

The first attempt to extract the Earth's heat from rocks with no pre-existing high permeability was the Fenton Hill HDR experiment. It was initially totally funded by the U.S. government, but later involved active collaborations under an International Energy Agency agreement with Great Britain, France, Germany, and Japan. The Fenton Hill site is characterized by a high-temperature-gradient, a large volume of uniform, low-permeability, crystalline basement rock. It is located on the margin of a hydrothermal system in the Valles Caldera region of New Mexico, not far from the Los Alamos National Laboratory where the project was conceived.

The Fenton Hill experience demonstrated the technical feasibility of the HDR concept by 1980, but none of the testing carried out yielded all the performance characteristics required for a commercial-sized system (sufficient reservoir productivity, maintenance of flow rates with sufficiently low pumping pressures, high cost of drilling deep (> 3 km) wells in hard rock becoming the dominant economic component in low-gradient EGS resources).

Country	Site	Depth (m)	Temperature (°C)	Dates	Production of electricity
France	Le Mayet		22 (production)	1975-1989	
	Soultz-sous-Forêts	5000 (3 boreholes)	155 (production)	1985-	1.5 MWe Since June 2008
Germany	Bad Urach	4500 and 2600	180 (reservoir)	1976-	
	Falkenberg	300	13 (reservoir)	1975-1985	
Switzerland	Basel	4500	180 (reservoir)		
UK	Rosemanowes	2700	70 (production)	1975-1991	
Australia	Cooper Basin	4250	212 (production)	2003-	
USA	Fenton Hill	5000	191 (production)	1973-2000	
	Desert Peak	5420		2001-	
Japan	Hijori	2300	180 (production)	1981-1987	

MWe : MWelectricity (raw production minus consumption of electricty required for the production).
Data after MIT (2006), Davatzes and Hickmann (2009), Genter et al. (2009) and Wyborn (2011).
Temperatures are given for the production phase for successful EGS sites or for the reservoir when no production occurred.

Table 6. Overview of some HDR or EGS programs in the world.

The program was divided into two major phases. Phase I (1974 – 1980), focused on a 3 km deep reservoir with a temperature of about 200°C. Phase II (1979-1995) penetrated into a deeper (4.4 km), hotter (300°C) reservoir. The two separate, confined HDR reservoirs were created by hydraulic fracturing and were flow-tested for almost a year each. A major lesson learned from the Fenton Hill HDR experience is that the characteristics of the joint system are highly variable : the joint-extension pressure in the Phase I reservoir was only half that obtained for the Phase II reservoir (MIT, 2006). This pressure is controlled by the interconnected joint structure that cannot be discerned either from borehole observations or from the surface. Only microseismic observations might show the portion of the induced seismicity that is really related to the opening of the joints allowing the main flow paths. However, by the early 1980s, HDR projects (Table 6) showed that in most of the cases, hydraulic stimulation did not only create new fractures but also re-opened by shearing natural joints favourably aligned with the principal directions of the local stress field and generally sealed by mineral deposits.

Several lessons were learnt at Fenton Hill. First, deep (5 km) high-temperature (up to 300 °C) wells can be completed in hard, abrasive rock. Second, it was possible to create or reactivate large-scale fracture networks and thanks to seismic monitoring and directed boreholes to intercept them. It was also possible to circulate the fractures with fluids thanks to the boreholes. The first models of flow and heat transfer were developed and used to predict the behaviour of the EGS reservoir. However, if injection pressures were lowered to reduce water loss and reservoir growth, the flow rates were lower than expected. An expert panel of the Massachusetts Institute of Technology estimated in 2006 that EGS could provide up to 100 000 megawatts of electricity in the United States by 2050, or about 10% of the current national capacity (high proportion for an alternative energy source). Up to US$132.9 million from the recovery act are to be directed at EGS demonstration projects.

4.2 Rosemanowes (UK; after MIT, 2006)

As a result of experience during Phase I at Fenton Hill, the Camborne School of Mines undertook an experimental HDR project at Rosemanowes (Cornwall, U.K.) in a granite. The project was funded by the U.K. Department of Energy and by the Commission of the European Communities. The temperature was restricted deliberately to below 100°C, to minimize instrumentation problems. This project was never intended as an energy producer but was conceived as a large-scale rock mechanics experiment about the stimulation of fracture networks. The site was chosen because of its clearly defined vertical jointing, high-temperature gradients between 30-40°C/km and its strike-slip tectonic regime.

Phase 1 of the project started in 1977, with the drilling of several 300 m wells dedicated to test fracture-initiation techniques. Phase 2 was characterized by the drilling of 2 wells that reached 2000 m and a temperature of nearly 80°C. Both were deviated in the same plane to an angle of 30 degrees from the vertical in the lower sections, and separated by 300 m vertically. Stimulation of the injection well was performed, initially with explosives, and then hydraulically at rates up to 100 kg/s and wellhead pressures of 14 MPa. A short circuit unfortunately developed between the two wells, which allowed cool injected water to return too rapidly to the production well: the temperature dropped from 80°C to 70°C. In phase 3A, with no further drilling, lowering the pressure in the production well seemed to close the joint apertures close to the borehole and increase the impedance. An experiment to place a proppant material (sand) in the joints near the production borehole was performed with a high viscosity gel and significantly reduced the water losses and impedance but also worsened the short circuiting and lowered the flow temperature in the production borehole even further. It was concluded that the proppant technique would need to be used with caution in any attempt to manipulate HDR systems. At Rosemanowes, it became clear that everything one does to pressurize a reservoir is irreversible and not necessarily useful for heat mining. For example, pumping too long at too high a pressure might cause irreversible rock movements that could drive short circuits as well as pathways for water losses to the far field (MIT, 2006). A packer assembly was placed close to the bottom of the borehole to seal off the short-cut and was successful but resulted in a subsequent low flow rate. This was interpreted as a new stimulated zone poorly connected to the previous one and demonstrated that individual fractures can have independent connections to the far-field fracture system leading to a globally poor connection of the reservoir.

4.3 Hijori (Japan; after MIT, 2006 and Yanagisawa et al, 2011)

This HDR project is located on Honshu island, on the edge of the Hijori caldera, where the high thermal gradient is related to a recent volcanic event (10 000 years old). The stress regime is very complex. The site was first drilled in 1989 after the results obtained at Fenton Hill to which Japan contributed. One injector and three producer wells were drilled from 1989 to 1991 between 1550 and 2151m. The temperature reached more than 225°C at 1500 m and 250°C at 1800 m. The spacing between the bottom of wells was about 40-55 m. The deep reservoir (about 2200 m), drilled from 1991 to 1995, was characterized by natural fractures. The distance between the wells, at that depth was 80 to 130 m. Hydraulic fracturing experiments began with injection of 2000 m³ of water. The stimulation was carried out in four stages at rates of 1, 2, 4 and 6 m³/min. A 30-day circulation test was conducted following stimulation. A combination of produced water and surface water was injected at

1-2 m³/min (17-34 kg/s), and steam and hot water were produced from 2 production wells. During the test, a total of 44500 m³ of water was injected while 13000 m³ of water were produced. The test showed a good hydraulic connection between the injector and the two producers, but more than 70% of the injected water was lost. The test was short and the reservoir continued to grow during the entire circulation period. After additional circulation tests in 1996, a one-year test began in 2000 for the shallow and the deep reservoirs with injection of 36°C water at 15-20 kg/s. Production of steam and water occurred at 4-5 kg/s at about 163-172°C. Total thermal power production was about 8 MWt. Test analysis showed that production was from both the deep and shallow reservoir. While the injection flow rate remained constant at about 16 kg/s, the pressure required to inject that flow decreased during the test from 84 to 70 bar. Total production from the two wells was 8.7 kg/s with a loss rate of 45%. Because of a dramatic cooling from 163°C to about 100°C, that long-term flow test was stopped. The measured change in temperature was larger than that predicted from numerical modelling. One lesson learnt from Hijori joined to Fenton Hill and Rosemanowes experiences was that it is better to drill a single well, stimulate it and map the acoustic emissions during stimulation, then drill additional wells into the acoustic emissions cloud rather than to try to drill two or more wells and attempt to connect them with stimulated fractures. In addition, injecting at low pressures for long time periods had an even more beneficial effect than injecting at high pressures for short periods. The Hijiori project also showed how important it is to understand not only the stress field but also the natural fracture system. Both Fenton Hill and Hijiori were on the edges of a volcanic caldera with very high temperature gradients (need for rather shallow wells, less expensive than deep ones) but also extremely complex parameters (geology, fractures, stress conditions) making these projects very challenging. The mineralogical composition of the Hijori EGS is close to that of Soultz and Habanero rock bodies and one can account for a rather similar chemical reaction with injected water, but the geological contexts are highly different resulting in different circulation schemes within the fracture networks.

4.4 Basel (Switzerland; after MIT, 2006 and Giardini, 2009)

Switzerland developed a Deep Heat Mining project to generate power and heat in Basel and Geneva. At Basel, in the southeastern end of the Rhine graben, close to the border with Germany and France, a 2.7 km exploration well was drilled, studied, and equipped with seismic instrumentation. A unique aspect of the Basel project is that drilling took place within city limits, and the heat produced by the system had the potential for cogeneration (direct use for local district heating as well as electricity generation). The project was initiated in 1996 and partly financed by the Federal Office of Energy together with private and public institutions. The plant was to be constructed in an industrial area of Basel, where the waste incineration of the municipal water purification plant provides an additional heat source. The core of the project, called Deep Heat Mining Basel, was a well triplet into hot granitic basement at a depth of 5 000 m. Two additional monitoring wells into the top of the basement rock were equipped with multiple seismic receiver arrays in order to record the fracture-induced seismic signals to map the seismic active domain of the stimulated reservoir volume. Reservoir temperature was expected to be 200°C. Water circulation of 100kg/s through one injection well and two production wells was designed to result in 30 MW of thermal power at wellheads. In combination with this heat source and an additional gas turbine, a combined cogeneration plant would have produced annually up to 108 GWh of electric power and 39 GWh of thermal power to the district heating grid. North-northwest trending compression and west-northwest extension creates a seismically active area into which the power plant was to be located. Therefore, it was important to record and understand the natural seismic activity as accurately as possible, prior to stimulation of a deep reservoir volume characteristically accompanied by induced seismicity. The first exploration well was drilled in 2001 into granitic basement at 2,650 m. The next well was planned to the targeted reservoir depth of 5000 m. On December 8th, 2006, an earthquake of magnitude 3.4 occurred, responsible for 7 million CHF of property damage. It has been attributed to stimulation operations. In such a seismically active area, one has also to consider the likely impact of the geothermal reservoir on the occurrence of a large earthquake like the event that caused large damage to the city in 1356. As a consequence of this 2006 earthquake, the Basel project was totally stopped in 2009. Many newspaper articles can be found about this story.

4.5 Habanero (Australia; after MIT, 2006 and Wyborn, 2011)

Australia has the hottest granites in the world thanks to radioactive decay characterized by temperatures approaching 250°C at a depth of 4 km in the Innamincka granite (Cooper Basin, south Australia) where the Habanero EGS is developed. Like at Soultz, the Habanero EGS is based on 3 drillings reaching a 4250 m depth. In this white two-mica granite containing 75%SiO_2, biotite is widely chloritized, feldspar is also altered and calcite precipitated as secondary mineral as already described for the Soultz granite (see section 3). Some fractures intersected in the first well were overpressured with water at 35 MPa above hydrostatic pressure. The fractures encountered were more permeable than expected likely because of slipping improving their permeability and resulting in drilling fluids being lost into them. The well intersected granite at 3668 m and was completed with a 6-inch open hole. It was stimulated in November and December 2003. A volume of 20000 cubic meters of water was injected into the fractures at flow rates from 13.5 kg/s to 26 kg/s, at pressures up to about 70 MPa. As a result, a volume estimated from acoustic emission data at 0.7 km^3 was developed into the granite body. A second well was drilled 500 m from the first one and intersected the fractured reservoir at 4325 m. During drilling pressure changes were recorded in the first well. The second well was tested in 2005 with flows up to 25kg/s and a surface temperature of 210°C was achieved. Testing between the two wells was delayed because of lost equipment in the second well. The first well was stimulated again with 20000 m^3 of water and it appeared, thanks to acoustic emission, that the old reservoir was extended by another 50% and finally covered an area of 4 km^2. A third well was drilled 568 m from the first one and was stimulated in 2008. The well productivity was doubled. As a result of these stimulations, two parallel fracture planes with a 15°W dip developed separated by about 100 m around 4200-4400 m and 4300-4600 m depths. The open-loop test performed in 2008 injected 18.5 kg/s in the first well. The third well produced 20 kg/s of water at a temperature around 212°C thanks to flow in the main fracture plane cited before. The productivity obtained during this test was nearly similar to that obtained in the Soultz GPK2 well allowing electricity production from June 2008. The main challenges to future progress are the reduction of drilling costs, an increased rate of penetration for drillings in hard formations, increasing flow rate by improving well connection to reservoir and through development of multiple reservoirs

(Wyborn, 2011). The concept of a 25MWe commercial plant is now designed with 3 injection wells and 6 production wells. The ultimate potential is to supply up to 6500 MWe of long-term base-loadpower, equivalent to electrical supply from ~750 MT thermal coal (Wyborn, 2011).

5. Forthcoming developments and challenges of EGS projects

Many research teams are currently working on improvement of existing techniques of innovation developments to ensure better production rates and minimized constraints. Among these innovations, the following are particularly promising but of course non exhaustive.

5.1 CO_2 EGS

The use of supercritical CO_2 as a heat transfer fluid has been first proposed as an alternative to water for both reservoir creation and heat extraction in EGS (Brown, 2000). Numerical simulations have shown that under expected EGS operating conditions, CO_2 could achieve more efficient heat extraction performance than water (Magliocco et al., 2011). CO_2 has numerous advantages for EGS: greater power output, minimized parasitic losses from pumping and cooling, carbon sequestration and minimized water use. Magliocco et al. (2011) have performed laboratory tests of CO_2 injection while Plaskina et al. (2011) made a numerical simulation study of effects of CO_2 injection to provide a new method to improve heat recovery from the geopressured aquifers by combining the effects of natural and forced convection.

5.2 *In situ* formation of calcium carbonate as a diversion agent

During stimulation of EGS wells, water is injected in order to open sealed fractures through shear failure. When the fractures are open, the stimulation fluid flows into them and becomes unavailable for stimulation elsewhere. Fluid diversion agents can serve to temporarily plug newly stimulated fractures in order to make the injected water available to stimulate new fractures (e.g. Petty et al., 2011). The diversion agent is subsequently removed to allow flow from those previously sealed fractures. As demonstrated by Ledésert et al. (2009) and Hébert et al., (2010), calcite is found naturally in fractures of EGS reservoirs and prevents the fluid from flowing into fractures. The in situ precipitation of calcium carbonate was studied by Rose et al. (2010) for use as a diversion agent in EGS.

5.3 Use of oil and gas reservoirs for EGS purposes

A lot of oil and gas reservoirs have been or will be abandoned in petroleum industry. According to Li and Zhang (2008) these oil and gas reservoirs might be transferred into exceptional enhanced geothermal reservoirs with very high temperatures. Air may be injected in these abandoned hydrocarbon reservoirs and in-situ combustion will occur through oxidization. The efficiency of power generation using the fluids from in-situ combustion reservoirs might be much higher than that obtained by using hot fluids coproduced from oil and gas reservoirs because of the high temperature.

6. Conclusion

Enhanced Geothermal Systems experiences at Fenton Hill (USA), Rosemanowes (UK), Hijori (Japan) and Basel (Switzerland) allowed scientists to develop a European thermal pilot-plant producing electricity in Soultz-sous-Forêts (France) since June 2008. This project is the result of 20 years of active research based on geology (petrography, mineralogy, fracture analysis), geochemistry, geophysics (seismic monitoring, well-logging), hydraulics and modelling. Technical improvements were also necessary to allow deep drilling (down to 5000 m) in a hard (granite), highly fractured rock and circulation of water at great depth (between 4500 and 5000 m). The rock behaves as a heat exchanger in which cold water is injected. The water circulates in the re-activated fracture planes where it warms up. It is pumped to the surface and activates a 1.5 MWe geothermal Organic Rankine Cycle power plant that converts the thermal energy into electricity. In such a project challenges are numerous and difficult since the injected water must circulate at great depths between the 3 wells of the triplet with no or little loss and the flow rate and fluid temperature must be and remain high enough to allow production of electricity. Provided careful monitoring of the reservoir during operation, EGS are a sustainable, renewable and clean way to produce electricity. It has been proven that environmental impacts of EGS are lower than those of nuclear or fossil fuel power plants dedicated to the production of electricity. The Soultz EGS pilot plant is the first one in the world to produce electricity and it should be followed in the forthcoming years by industrial units that will produce electricity at a commercial scale. Many other EGS projects have begun all around the world and a lot of scientific and technical targets are in development to improve the production of energy (electricity and central heating through district networks).

7. References

Baisch, S., Carbon, D., Dannwolf, U., Delacou, B., Devaux, M., Dunand, F., Jung, R., Koller, M., Matin, C., Sartori, M., Secanelll, R. & Uörös, R. (2009). Deep Heat Mining – Seismic Risk Analysis, *Departement für Wirtschaft, Soziales und Umwelt des Kantons Basel-Stadt, Armt für Umwelt und Energie*, Available at http://www.wsu.bs.ch/serianex_teil_1_english.pdf

Bartier, D., Ledésert, B., Clauer, N., Meunier, A., Liewig, N., Morvan, G, Addad, A. (2008). Hydrothermal alteration of the Soultz-sous-Forêts granite (Hot Fractured Rock geothermal exchanger) into a tosudite and illite assemblage, *European Journal of Mineralogy*, 20 (1), 131-142.

Beardsmore, G.R. & Cooper, G.T. (2009). Geothermal Systems Assesment- Identification and Mitigation of EGS Exploration Risk, *Proceedings of the Thirty-Fourth Workshop on Geothermal Reservoir Engineering*, Stanford University, Stanford, California, February 9-11, 2009, SGP-TR-187

Brown, D.W. (2000). A Hot Dry Rock Geothermal Energy Concept Utilizing Supercritical CO2 Instead of Water, *Proceedings of the Twenty-Fifth Workshop on Geothermal Reservoir Engineering*, Stanford University, Stanford, California, January 24–26, 2000, SGP-TR-165.

Brown, D.W. (2009). Hot Dry Rock Geothermal Energy : Important Lessons from Fenton Hill, *Proceedings of the Thirty-Fourth Workshop on Geothermal Reservoir Engineering*, Stanford University, Stanford, California, February 9-11, 2009, SGP-TR-187

Clauser, C. (2006), Geothermal Energy, in K. Heinloth (ed.), Landolt-Börnstein, Group VIII *Advanced Material and Technologies*, Vol. 3 Energy Technologies, Subvolume C, Renewable Energies, 480–595, Springer Verlag, Heidelberg-Berlin.

Concha, D., Fehler, M., Zhang, H. & Wang, P. (2010). Imaging of the Soultz Enhanced Geothermal Reservoir Using Microseismic Data, *Proceedings of the Thirty-Fifth Workshop on Geothermal Reservoir Engineering*, Stanford University, Stanford, California, February 1-3, 2010, SGP-TR-188

Culver, G. (1998) Drilling and Well Construction, Chapter 6, *Geothermal Direct Use Engineering and Design Guidebook*, Third Edition, Culver, Gene, Geo-Heat Center, Oregon Institute of Technology, Klamath Falls, OR, 1998. http://geoheat.oit.edu/pdf/tp65.pdf

Davatzes, N.C. & Hickman, S.H. (2009). Fractures, Stress and Fluid Flow prior to Stimulation of well 27-15, Desert Peak, Nevada, EGS Project, Clark,C. (2009). Pre-production Activity Impacts of Enhanced Geothermal Systems, *Proceedings of the Thirty-Fourth Workshop on Geothermal Reservoir Engineering*, Stanford University, Stanford, California, February 9-11, 2009, SGP-TR-187

Dezayes, C. and Genter, A. (2008). Large-scale Fracture Network Based on Soultz Borehole Data, EHDRA Scientific Conference, *Proceedings of the EHDRA scientific conference* ,24–25 September 2008, Soultz-sous-Forêts, France.

Dezayes C., Genter A., Valley B. (2010). Structure of the low permeable naturally fractured geothermal reservoir at Soultz, *Comptes Rendus Geosciences*, 342, 517-530.

Dèzes, P., Schmid, S.M., Ziegler, P.A. (2004). Evolution of the European Cenozoic Rift System: interaction of the Alpine and Pyrenean orogens with their foreland lithosphere, *Tectonophysics* , 389, 1–33.

Dubois, M., Ledésert, B., Potdevin, J.L. & Vançon, S. (2000). Détermination des conditions de précipitation des carbonates dans une zone d'altération du granite de Soultz (soubassement du fossé rhénan, France) : l'enregistrement des inclusions fluides, *Comptes Rendus de l'Académie des Sciences*, 331, 303-309.

Economides, M.J. & Nolte, K.G. (2000). *Reservoir Stimulation*, Third Edition, Wiley, NY and Chichester

Fridleifsson, I.B. (2000). Improving the Standard of Living, World Geothermal Congress 2000 Convention News 2, 1.

Genter, A., Traineau, H., Dezayes, C., Elsass, P., Ledésert, B. , Meunier, A. & Villemin. T. (1995). Fracture Analysis and Reservoir Characterization of the Granitic Basement in the HDR Soultz Project (France), *Geothermal Science and Technology*, 4(3), 189-214.

Genter, A., Fritsch, D., Cuenot, N., Baumgärtner, J. & Graff J.J. (2009) Overview of the Current Activities of the European EGS Soultz Project : From Exploration to Electricity Production, *Proceedings of the Thirty-Fourth Workshop on Geothermal Reservoir Engineering*, Stanford University, Stanford, California, February 9-11, 2009, SGP-TR-187

Genter, A., Cuenot, N., Goerke, X. & Sanjuan B., (2010). Programme de suivi scientifique et technique de la centrale géothermique de Soultz pendant l'exploitation, Rapport d'avancement Phase III : activité 2010, décembre 2010, rapport GEIE EMC RA05 001 P, 66 pp.

Gérard, A. & Kappelmeyer, O. (1987). The Soultz-sous-Forêts project : Proceedings of the first EEC/US workshop on geothermal Hot dryRocks Technology, *Geothermics*, 393-399.

Gérard, A., Genter, A., Kohl, Th., Lutz, Ph., Rose, P. & Rummel, F. (2006). The deep EGS (Enhanced Geothermal System) Project at Soultz-sous-Forêts (Alsace, France), *Geothermics*, 35, No. 5-6, 473-483.

Giardini, D. (2009). Geothermal quake risks must be faced, *Nature*, 462, 848-849

Hébert, R.L., Ledésert, B., Bartier, D., Dezayes, C., Genter, A., & Grall, C. (2010) The Enhanced Geothermal System of Soultz-sous-Forêts: A study of the relationships between fracture zones and calcite content, *Journal of Volcanology and Geothermal Research*, 196, 126–133.

Hébert, R.L, Ledésert, B. , Genter, A., Bartier, D. & Dezayes, C. (2011) Mineral Precipitation in Geothermal Reservoir : the Study Case of Calcite in the Soultz-sous-Forêts Enhanced Geothermal System, *Proceedings of the Thirty-Sixth Workshop on Geothermal Reservoir Engineering*, Stanford University, Stanford, California, January 31 - February 2, 2011, SGP-TR-191

Hickman, S. & Davatzes, N. (2010).In-Situ Stress and Fracture Characterization for Planning of an EGS Stimulation in the Desert Peak Geothermal Field, NV; *Proceedings of the Thirty-Fifth Workshop on Geothermal Reservoir Engineering*, Stanford University, Stanford, California, February 1-3, 2010, SGP-TR-188

Hurtig, E., Cermak, V., Haenel, R. Zui, V. (1992). Geothermal Altlas in Europe, *Hermann Haak Verlagsgeschellschaft mbH*, Germany.

Karvounis D. & Jenny. P. (2011). Modeling of Flow and Transport in Enhanced Geothermal Systems, *Proceedings of the Thirty-Sixth Workshop on Geothermal Reservoir Engineering*, Stanford University, Stanford, California, January 31 - February 2, 2011, SGP-TR-191

Kosack, C., Vogt, C., Marquart, G., Clauser, C. & Rath, V. (2011). Stochastic Permeability Estimation for the Soultz-sous-Forêts EGS Reservoir, *Thirty-Sixth Workshop on Geothermal Reservoir Engineering Stanford University*, Stanford, California, January 31 - February 2, 2011, SGP-TR-191

Kretz, R. (1985). Symbols for Rock- forming Minerals , American Mineralogist, 68, 277-279.

Ledésert, B., Dubois, J., Genter, A. & Meunier, A. (1993) Fractal analysis of fractures applied to Soultz-sous-Forêts Hot Dry Rock geothermal program, *Journal of Volcanology and Geothermal Research*, 57, 1-17.

Ledésert, B., Berger, G., Meunier, A., Genter, A. and Bouchet, A. (1999). Diagenetic-type reactions related to hydrothermal alteration in the Soultz-sous-Forêts granite, *European Journal of Mineralogy*, 11, 731-741.

Ledésert B., Hébert R., Grall C., Genter A., Dezayes C., Bartier D., Gérard A. (2009) Calcimetry as a useful tool for a better knowledge of flow pathways in the Soultz-sous-Forêts Enhanced Geothermal System, *Journal of Volcanology and Geothermal Research*, 181, 106-114.

Ledésert, B., Hébert, R., Genter, A., Bartier, D., Clauer ,N. & Grall C. (2010) Fractures, Hydrothermal alterations and permeability in the Soultz Enhanced Geothermal System, *Comptes Rendus Géosciences*, 342 (2010) 607–615.

Li, K. & Zhang, L. (2008). Exceptional enhanced geothermal systems from oil and gas reservoirs, *Proceedings of the Thirty-Third Workshop on Geothermal Reservoir Engineering*, Stanford University, Stanford, California, January 28-30, 2008, SGP-TR-185

Lund, J.W. (2007). Characteristics, Development and Utilization of Geothermal Resources, *Geo-Heat Center Quarterly Bulletin*, Vol. 28, No. 2, Geo-Heat Center, Oregon Institute of Technology, Klamath Falls, Or, Available from http://geoheat.oit.edu/pdf/tp126.pdf

Magliocco, M., Kneafsey, T.J., Pruess, K. & Glaser, S. (2011) Laboratory Experimental Study of Heat Extraction from Porous Media by Means of CO_2, *Proceedings of the Thirty-Sixth Workshop on Geothermal Reservoir Engineering*, Stanford University, Stanford, California, January 31 - February 2, 2011, SGP-TR-191

Minissale, A. (1991). The Larderello Geothermal Field: a Review, *Earth-Science Reviews*, Volume 31, Issue 2, Pages 133-151

M.I.T (2006). The Future of Geothermal Energy: Impact of Enhanced Geothermal Systems (EGS) on the United States in the 21st Century, *Massachusetts Institute of Technology*, ISBN: 0-615-13438-6, Available from http://geothermal.inel.gov.

Nami, P., Schellschmidt, R., Schindler, M. & Tischner, T. (2008) Chemical Stimulation Operations for Reservoir Development of the deep Crystalline HDR/EGS at Soultz-sous-Forêts (France), *Proceedings of the Thirty-Third Workshop on Geothermal Reservoir Engineering*, Stanford University, Stanford, California, January 28-30, 2008, SGP-TR-185

Petty, S., Bour, D., Nordin, Y., Nofziger, L. (2011) Fluid Diversion in an Open Hole Slotted Liner – a First Step in Multiple Zone EGS stimulation, *Proceedings of the Thirty-Sixth Workshop on Geothermal Reservoir Engineering*, Stanford University, Stanford, California, January 31 - February 2, 2011, SGP-TR-191

Plaksina, T., White, C., Nunn, J. & Gray, T. (2011). Effects of Coupled Convection and CO2 Injection in Stimulation of Geopressured Geothermal Reservoirs, *Proceedings of the Thirty-Sixth Workshop on Geothermal Reservoir Engineering*, Stanford University, Stanford, California, January 31 - February 2, 2011, SGP-TR-191

Portier, S., Vuataz, F.D., Nami, P., Sanjuan, B. & André Gérard, A. (2009). Chemical Stimulation Techniques for Geothermal Wells: Experiments on the Three-well EGS System at Soultz-sous-Forêts, France, *Geothermics*, 38, 349–359

Radilla, G., Sausse, J., Sanjuan, B. & Fourar, M. (2010). Tracer Tests in the Enhanced Geothermal System of Soultz-sous-Forêts. What Does the Stratified Medium Approach Tells us about the Fracture Permeability in the Reservoir ? *Proceedings of the Thirty-Fifth Workshop on Geothermal Reservoir Engineering*, Stanford University, Stanford, California, February 1-3, 2010, SGP-TR-188

Rafferty, K. (1998). Outline Specifications for Direct-Use Wells and Equipment, , Geo-Heat Center, Oregon Institute of Technology, Klamath Falls, OR, 1998. http://geoheat.oit.edu/pdf/tp66.pdf

Romagnoli, P., Arias, A., Barelli, A., Cei, M. & Casini, M. (2010). An Updated Numerical Model of the Larderello–Travale Geothermal System, Italy, *Geothermics*, Volume 39, Issue 4, Pages 292-313

Sanjuan, B., Pinault, J-L, Rose, P., Gérard, A., Brach, M., Braibant, G., Crouzet, C., Foucher, J-C, Gautier, A. & Touzelet S. (2006). Tracer Testing of the Geothermal Heat Exchanger at Soultz-sous-Forêts (France) between 2000 and 2005, *Geothermics*, 35, No. 5-6, 622-653.

Sanval, S.K. & Enedy, S. (2011) Fifty Years of Power Generation at the Geysers Geothermal Field, California – The Lessons Learned, *Proceedings of the Thirty-Sixth Workshop on Geothermal Reservoir Engineering*, Stanford University, Stanford, California, January 31 - February 2, 2011, SGP-TR-191

Satman, A. (2011). Sustainability of Geothermal Doublets, *Proceedings of the Thirty-Sixth Workshop on Geothermal Reservoir Engineering*, Stanford University, Stanford, California, January 31 - February 2, 2011, SGP-TR-191

Sausse, J., Dezayes, C., Genter, A. (2007). From geological interpretation and 3D modelling to the characterization of deep seated EGS reservoir of Soultz (France), Proceedings European Geothermal Congress 2007, Unterhaching, Germany, 30 May–1 June 2007, 7 pp.

Sausse, J., Dezayes, C., Genter, A. & Bisset, A. (2008). Characterization of Fracture Connectivity and Fluid Flow Pathways Derived from Geological Interpretation and Modelling of the Deep Seated EGS Reservoir of Soultz (France), Proceedings of the Thirty-Third Workshop on Geothermal Reservoir Engineering Stanford University, Stanford, California, January 28-30, 2008, SGP-TR-185

Sausse, J., Dezayes, Ch., Dorbath, L., Genter, A. & Place, J. (2009). 3D Fracture Zone Network at Soultz Based on Geological Data, Image Logs, Microseismic Events and VSP Results, *Geoscience*.

Scott, S., Gunnarsson, I. & Stef, A. (2011). Gas Chemistry of the Hellisheiði Geothermal Field, SW-Iceland, *Proceedings of the Thirty-Sixth Workshop on Geothermal Reservoir Engineering*, Stanford University, Stanford, California, January 31 - February 2, 2011, SGP-TR-191

SoultzNet (2011). http://www.geothermie-soultz.fr

Rose, P., Scott Fayer, S., Susan Petty, S. & Bour, D. (2010). The In-situ Formation of Calcium as a diversion Agent for Use in Engineered Geothermal Systems, Proceedings of the Thirty-Fifth Workshop on Geothermal Reservoir Engineering Stanford University, Stanford, California, February 1-3, 2010, SGP-TR-188

Tamanyu, S., Fujiwara, S., Ishikawa, J.I. & Jingu, H. (1998). Fracture System Related to Geothermal Reservoir Based on Core Samples of Slim Holes. Example from the Uenotai Geothermal Field, Northern Honshu, Japan, *Geothermics*, Volume 27, Issue 2, April 1998, Pages 143-166

Wyborn, D. (2011). Hydraulic Stimulation of the Habanero Enhanced Geothermal System (EGS), South Australia, 5th BC Unconventional Gas Technical Forum April 2011, Geodynamics Limited, available from
http://www.em.gov.bc.ca/OG/oilandgas/petroleumgeology/UnconventionalGas/Documents/2011Documents/D%20Wyborn.pdf

Yanagisawa, N., Matsunaga, I., Ngothai, Y. & Wyborn, D. (2011) Geochemistry Change during Circulation Test of EGS Systems, *Proceedings of the Thirty-Sixth Workshop on Geothermal Reservoir Engineering,* Stanford University, Stanford, California, January 31 - February 2, 2011, SGP-TR-191

Yasukawa, K. & Takasugi, S. (2003). Present Status of Underground Thermal Utilization in Japan, *Geothermics*, Volume 32, Issues 4-6, 609-618

Part 3

Fouling of Heat Exchangers

5

Self-Cleaning Fluidised Bed Heat Exchangers for Severely Fouling Liquids and Their Impact on Process Design

Dick G. Klaren and Eric F. Boer de
KLAREN BV
The Netherlands

1. Introduction

The invention of the self-cleaning fluidised bed heat exchangers dates back to 1971 when the principal author of this chapter was involved in the discovery and development of a very unique Multi-Stage Flash (MSF) evaporator for the desalination of seawater. The condensers used in this thermal desalination plant used stationary fluidised beds in multi-parallel condenser tubes. The particles fluidised in these tubes consisted of glass beads of 2 mm diameter. These small glass beads knocked of scale crystals from the tube wall at their very early stage of formation and, moreover, the turbulence created by the stirring action of the glass beads in the liquid caused thinning of the laminar boundary layer. This dramatically improved the heat transfer film coefficient in spite of very low liquid velocities in the tubes and reduced pumping power requirements.

Since the early 80s, the chemical processing industries showed a lot of interest for this unique heat exchanger, which seemed to be able to solve any fouling problem, even those problems, which required cleaning of conventional heat exchanger every few days or even hours.

In the next paragraphs we will pay attention to the consequences of heat exchanger fouling and in particular its cost. We explain the self-cleaning fluidised bed technology and also present a couple of installations. We also show some examples where the benefits of the self-cleaning fluidised bed heat exchange technology are responsible for a much wider range of advantages with respect to process design than non-fouling heat exchange only.

2. Fouling of heat exchangers

2.1 Consequences of heat exchanger fouling

It can be stated that a general solution to heat exchanger fouling still does not exist. This is not surprising, as knowledge of underlying mechanisms of the fouling process remains limited. Moreover, fouling in heat exchangers often concerns different types of heat exchangers, each with its own unique characteristics. Also, there are large differences in physical properties of the fluids to be applied in the exchangers. The consequences of heat exchanger fouling are:

- Loss of energy,
- loss of production or reduced capacity operation,
- over sizing and / or redundancy of equipment,
- excessive maintenance cost,
- hazardous cleaning solution handling and disposal.

Over sizing of heat transfer equipment has become an accepted approach to increase the period of time necessary to reach the fouled state. The equipment is then cleaned (chemically or mechanically) to return the heat transfer surface to a near clean condition with recurring maintenance cost and the possibility of cleaning solution disposal problems. Later in this chapter, it will be shown that there are existing cases where over sizing of heat transfer surface can involve the installation of two to five times the surface required for the clean condition. Also, in these severe cases it may be necessary to carry out the cleaning procedure every two or three days, resulting in excessive downtime, maintenance costs and solution disposal problems. Sometimes the fouling problems are so severe that heat transfer performance reduces to almost zero in a matter of hours.

Experience has shown that the alternatives to recurring fouling problems associated with the cooling or heating of a severe fouling liquid are certainly limited. In the case of cooling applications unsuccessful attempts to recover energy from hot waste streams may lead to the total abandonment of an otherwise promising energy management program.

Frequently the only acceptable approach to heating severe fouling liquids will involve direct steam injection. This results in a loss of condensate and the dilution of the process stream, which often requires costly reconcentration later in the process. However, heating by direct steam injection does offer a unique opportunity to define the actual cost of fouling in terms of lost condensate and the subsequent cost of water removal. In the next paragraph we will pay attention to the very high cost of heat exchanger fouling on a global scale, and for one process in particular.

2.2 Cost associated with heat exchanger fouling

The heat exchangers in a crude oil train of a refinery for the distillation of crude oil in lighter fractions are often subject to severe fouling, and do represent globally a very high level of cost. In this sub-paragraph we like to explain this particular example in a nutshell. For a much more detailed explanation one is referred to Ref. [5].

Fig. 1 gives an schematic impression of the heating of crude oil in a crude oil train downstream the desalter and in the furnace, where after the oil is cracked in much lighter fractions in the distillation column. Fig. 2 gives an impression of the temperatures in this very much simplified example.

Fouling of the crude oil heat exchangers downstream the desalter, Ex4 up to and inclusive Ex8, is shown at first instance by a drop of the inlet temperature 271 °C of the crude oil in the furnace, which means that more heat has to be supplied into the furnace to meet the required outlet temperature 380 °C of the crude oil entering the distillation column. This, of course, a phenomenon caused by fouling of the heat exchangers, does requires extra fuel (i.e. extra energy) to be burned in the furnace to keep the distillation facility in operation. At a certain moment, the inlet temperature of the crude oil in the furnace has dropped to such an

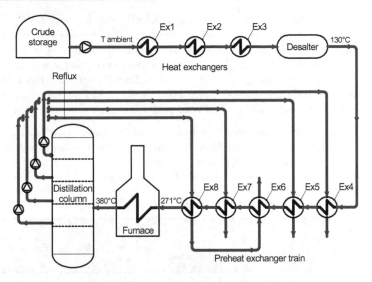

Fig. 1. Simplified flow diagram of a crude oil preheat train.

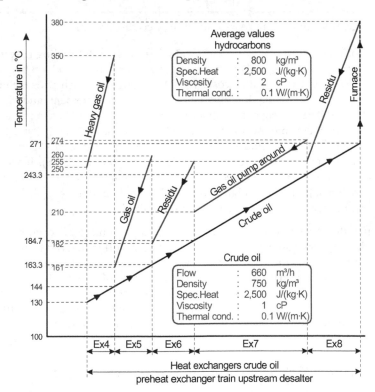

Fig. 2. Temperature diagram crude oil preheaters in simplified flow diagram shown in Fig. 1.

extent that the heating capacity of the furnace is insufficient to meet the required outlet temperature of the crude oil. This temperature can only be maintained by reducing the throughput of crude oil through the heat exchanger train, which, however, also reduces the production capacity of the refinery. This example shows very clearly that fouling of heat exchangers does cost extra energy and may also reduce the production capacity of an installation. For our crude oil preheat train, both facts, including the maintenance cost, increases the refining cost for each barrel of crude oil. What are these costs on a global scale?

At this moment (2011), the global production of crude oil amounts to approx. 85 million barrels per day (bpd). Table 1 has been derived from information given in Ref. [5], and gives an impression about the annual fouling cost for the crude oil being processed in the crude oil preheater trains of all refineries in the world as a function of the price per barrel crude oil.

	Crude oil price in US$ per barrel			
	$ 45	$ 60	$ 75	$ 90
Fouling costs in billion US dollars	10.9	12.5	14.1	15.7

Table 1. Fouling costs crude oil trains as a function of crude oil price.

It is assumed that for a crude oil price of US$ 60 / barrel, the total fouling cost in crude oil preheat trains processing the global crude oil production of 85 million bpd represents approx. 10 % of the worldwide fouling costs in heat exchangers, which costs include all kind of heat exchangers for both liquids and gases. From this statement and the numbers presented in Table 1, it can be concluded that the total cost the world has to pay annually for fouling of heat exchangers amounts to approx. US$ 125 billion. In Ref. [1], Garrett-Price used a different approach and concluded that the fouling of heat exchangers do cost an industrialised nation approx. 0.3 % of its Gross National Product (GNP). If we apply this rule to the GNP of the whole world (2007) of US$ 55 000 billion, then we find for the global fouling cost US$ 165 billion. This is higher than US$ 125 billion, but, very likely, because not all countries can be considered as sufficiently industrialised.

It is evident that the often excessive costs of heat exchanger fouling have led to a number of initiatives to develop some additional alternative solutions, often derived from research into the various fouling mechanisms. Over a period of forty years, the principal author Dr. Ir. Dick G. Klaren has participated in the development of one of the more promising alternatives: The self-cleaning or non-fouling fluidised bed heat exchanger. During this period the concept was taken from a laboratory tool to a fully developed heat transfer tool, which is now used to resolve severe fouling problems in a range of applications throughout the process industries.

3. Principle of the self-cleaning fluidised bed heat exchanger

Over the past 40 years, the principle of the fluidised bed heat exchange technology evolved from a type that applied a stationary fluidised bed into a more widely applicable concept that uses a circulating fluidised bed. This section pays attention to both principles of which the circulating concept is more widely applicable in comparison with the stationary type.

In principle such a stationary fluidised bed heat exchanger consists of a large number of parallel vertical tubes, in which small solid particles are kept in a stationary fluidised condition by the liquid passing up the tubes. The solid particles regularly break through the boundary layer of the liquid in the tubes, so that good heat transfer is achieved in spite of comparatively low liquid velocities in the tubes. Further, the solid particles have a slightly abrasive effect on the tube wall of the exchanger tubes, removing any deposit at an early stage.

Fig. 3 shows a heat exchanger with a stationary fluidised bed, which means there is no change in position of the particles as a function of time. The inlet channel contains a fluidised bed and a flow distribution system which is of utmost importance to achieve stable operation of all parallel exchanger tubes, or said otherwise: Equal distribution of liquid and solid particles over all the tubes. This exchanger is characterised by the use of glass beads with diameters of 2 to 3 mm and very low liquid velocities in the tubes. The glass beads are fluidised along the tubes and form a shallow fluidised bed layer in the outlet channel. This exchanger is only suitable for operation on constant flow.

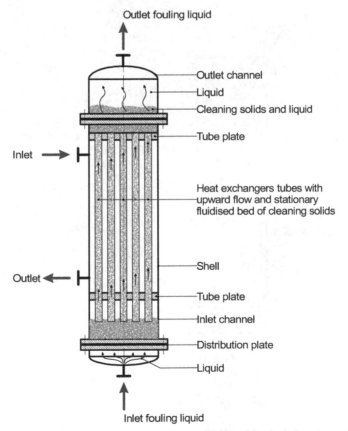

Fig. 3. Self-cleaning heat exchanger with stationary fluidised bed of cleaning solids.

Fig. 4 shows a heat exchanger with an 'internally circulating' fluidised bed. In this heat exchanger the liquid and particles flow through the tubes from the inlet channel into the widened outlet channel, where the particles disengage from the liquid and are returned to the inlet channel through multiple downcomer tubes, which are uniformly distributed over the actual heat exchanger or riser tubes. Now, the particles in the tubes experience a change of position with time. This heat exchanger can also use higher density materials like chopped metal wire as particles with dimensions up to 4 mm, and normally operates on higher liquid velocities in the tubes than the exchanger with the stationary fluidised bed. Depending on the design, this exchanger can also operate on a varying flow and in case of chopped metal wire particles; this exchanger represents the ultimate tool for handling the most severe fouling problems in liquid heat transfer.

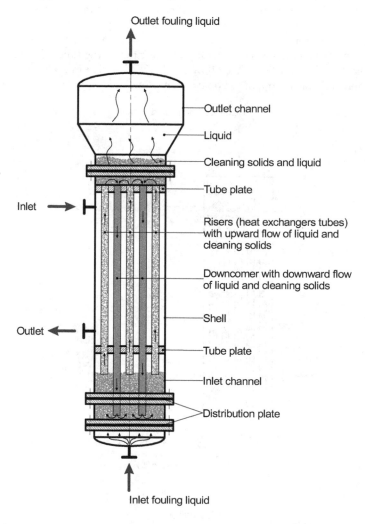

Fig. 4. Self-cleaning heat exchanger with internal circulation of cleaning solids.

The heat exchanger shown in Fig. 5 applies an 'externally circulating' fluidised bed. In this heat exchanger the liquid en particles flow from the outlet channel into an external separator where the particles are separated from the liquid, where after the particles flow from the separator into the inlet channel through only one downcomer and control channel For hydraulic stability reasons, this heat exchanger has the advantage that it only uses one downcomer, and the flow through this external and accessible downcomer can be monitored, influenced and varied by the control flow through line 1B. This flow only represents approximately 5 % of the feed flow through line 1 and shutting off this flow makes it possible to use the particles intermittently. This configuration also makes it possible to revamp existing severely fouling vertical conventional heat exchangers into a self-cleaning configuration as will be presented later in this chapter; this is a major advantage in comparison with the configuration shown in Fig. 4.

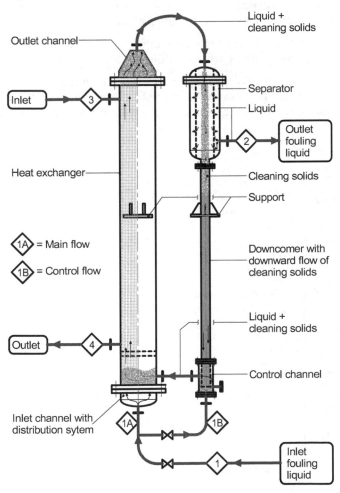

Fig. 5. Self-cleaning heat exchanger with external circulation of cleaning solids.

4. Performance of the self-cleaning fluidised bed heat exchanger

The performance of a self-cleaning fluidised bed heat exchanger and its design consequences have to be divided in the following subjects:

- Heat transfer correlation.
- Design consequences.
- Pumping power requirements.
- Fouling removal.
- Wear.

4.1 Heat transfer correlation

We briefly explain the composition of the tube-side heat transfer correlation for a heat exchanger which also applies recirculation of the particles and the liquid.

Fig. 6 shows the significant liquid velocities influencing the wall-to-liquid heat transfer coefficient for an exchanger with a circulating fluidised bed, such as:

U_s = superficial liquid velocity in the tubes relative to the tube wall,
$U_{b,w}$ = velocity of (moving) swarm of fluidised particles relative to the tube wall,
$U_{l,s}$ = superficial liquid velocity relative to the boundary limits of the (moving) swarm.

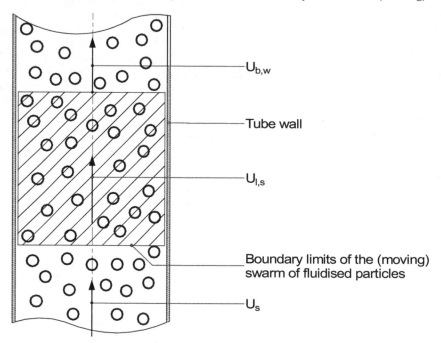

Fig. 6. Significant liquid velocities in tube of exchanger with circulating fluidised bed.

Where the superficial liquid velocity refers to the tube liquid velocity in the empty tube. From the explanation above, it follows:

$$U_s = U_{b,w} + U_{l,s} \tag{1}$$

For $U_{b,w} = 0$, the circulating fluidised bed satisfies the conditions of a stationary fluidised bed, which then yields:

$$U_s = U_{l,s} \tag{2}$$

where $U_{l,s}$ follows from the theory presented by Richardson and Zaki, Ref. [10].

The heat transfer coefficient $\alpha_{w,l}$ between the wall and the liquid of a circulating fluidised bed exchanger, is composed as follows:

$$\alpha_{w,l} = \alpha_l + \alpha_c \tag{3}$$

where:

α_l = wall-to-liquid heat transfer coefficient of a stationary fluidised bed with a superficial velocity $U_{l,s}$ related to the porosity ε of the bed
α_c = wall-to-liquid heat transfer coefficient for forced convection in a tube, taking into account a liquid velocity $U_{b,w}$, which actually corresponds with the velocity of the (stationary) fluidised bed moving along the tube wall

For the heat transfer coefficient α_l one is referred to Ruckenstein, Ref. [11], as long as superficial liquid velocities are calculated from porosities (ε) lower than 0.9. For porosities in the range $0.9 < \varepsilon \le 1.0$, the following equation is suggested:

$$\alpha_l = \alpha_l|_{\varepsilon=1.0} + \frac{(1-\varepsilon)}{(1-0.9)} \times \left\{ \alpha_l|_{\varepsilon=0.9} - \alpha_l|_{\varepsilon=1.0} \right\} \tag{4}$$

The heat transfer coefficient $\alpha_l|_{\varepsilon=1.0}$ is calculated using the equation of Dittus and Boelter taking into account the liquid velocity in the tube which corresponds with the terminal falling velocity on one single particle in the tube, i.e. $\varepsilon = 1.0$, as the liquid velocity used in the Reynolds number.

The heat transfer coefficient α_c is also obtained using the equation of Dittus and Boelter with $U_{b,w}$ as the liquid velocity used in the Reynolds number.

Fig. 7 shows the wall-to-liquid heat transfer coefficients in an exchanger with a circulating fluidised bed as a function of the various process parameters using 2.0 mm glass particles. It should be noticed that in Fig. 7 the curve $U_s = U_{l,s}$ shows the relation between heat transfer coefficient and relevant parameters for the stationary fluidised bed bed.

It should be emphasised that this heat transfer correlation is only an attempt to produce some approximate numbers for the overall heat transfer coefficients for any preliminary design. The real numbers which should be used in the performance guarantee of the heat exchanger follow from experimental operation of a representative pilot plant. Such a pilot plant is anyhow necessary to demonstrate the non-fouling operation.

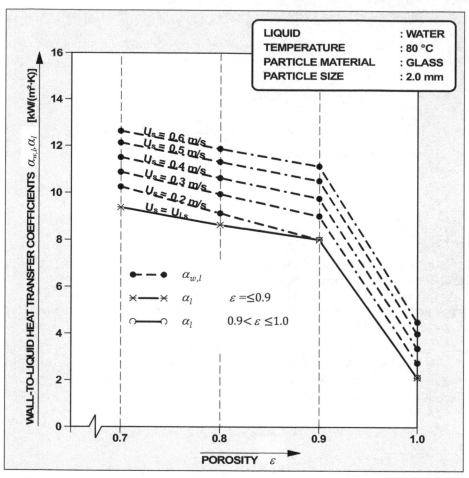

Fig. 7. Heat transfer coefficients in exchanger with circulating fluidised bed.

4.2 Design consequences

A fluidised bed exchanger offers the possibility to obtain heat transfer film coefficients at the tube-side of the same order of magnitude as normally achieved in conventional tubular exchangers, although at much lower liquid velocities. For example, a stationary fluidised bed heat exchanger using glass beads of only 2 mm with a porosity of the fluidised bed in the tubes of 75 % (i.e. the liquid volume fraction in the tube) can achieve heat transfer film coefficients of approx. 10 kW/(m² ·K) at a superficial velocity of approx. 0.12 m/s, which can only be realised in a conventional tubular heat exchanger with a liquid velocity of approx. 1.8 m/s.

The design consequences of this unique behaviour for a fluidised bed heat exchanger can be best explained with the help of the equation below for a heat exchange tube, which has been derived from the conservation equations for mass and energy:

$$L_t = D_o \times \left(\frac{D_i}{D_o}\right)^2 \times \frac{\rho_l \times c_l \times V_l}{4 \times k} \times \frac{\Delta T}{\Delta T_{\log}} \tag{5}$$

Where:

L_t	=	Tube length	[m]
D_o	=	Outer diameter of the tube[m]	
D_i	=	Inner diameter of the tube [m]	
ρ_l	=	Density of the liquid	[kg/m³]
c_l	=	Specific heat of the liquid	[J/(kg·K)]
V_l	=	Superficial liquid velocity in the tube	[m/s]
k	=	Overall heat transfer coefficient	[W/(m²·K)]
ΔT	=	Temperature difference of the liquid between tube inlet and tube outlet [°C]	
$\Delta T_{\log}$	=	Logarithmic mean temperature difference across tube [°C]	

For a comparison of the tube length between different types of exchangers for the same duty and temperatures, Equation (5) can be simplified:

$$L_t = C_1 \times \frac{D_o \times V_l}{k} \tag{6}$$

Where C_1 is a constant for a particular installation/application.

Or in words: The length of the tubes L_t is directly proportional to the diameter of the tube D_o, the liquid velocity in the tubes V_l, but inversely proportional to the heat transfer coefficient k. It can be stated that the clean k-values for self-cleaning fluidised bed heat exchangers are always somewhat higher than for the conventional heat exchangers at their normally much higher liquid velocities, with the remark that the clean k-values for the self-cleaning fluidised bed heat exchangers correspond with the design values used for the full-size self-cleaning fluidised bed heat exchanger with no need for cleaning, while for the conventional heat exchangers due to fouling the design k-value may be 2, 3, 4 or even 5 times lower than the clean k-value and frequent cleanings may be still necessary.

What the design consequences of excellent heat transfer at very low liquid velocities do mean for a self-cleaning fluidised bed heat exchanger in comparison with a conventional heat exchanger for the same application can be best explained with the following striking example:

A conventional Multi-Stage Flash (MSF) evaporator for seawater desalination with a seawater velocity of 1.8 m/s in the condenser tubes of 19.05 × 1.21 mm and an average heat transfer coefficient of 2500 W/(m²·K) required a total length of the condenser tubes of 173 m. Depending on the design of this seawater evaporator, this tube length requires the installation of 8 evaporator vessels in series, each vessel with a length of 20 m or even more.

The same MSF desalination plant equipped with stationary fluidised bed heat exchangers required a seawater velocity in the tubes of only 0.125 m/s to maintain a fluidised bed in all parallel operating tubes with a porosity of 75 % using glass particles with a density of

2750 kg/m³ and a diameter of 2 mm. In spite of this low seawater velocity, an overall heat transfer coefficient of 2500 W/(m²·K) was achieved. From the equations above, it can be concluded that this desalination plant required only 0.125 / 1.8 × 173 = 12 m condenser tube length in series, which can be installed in only one vessel with an overall height of less than 15 m.

4.3 Pumping power requirements

Pumping power is influenced by the pressure drop across the heat exchanger and the pressure drop to support a stationary fluidised bed, which is determined by the following equation:

$$\Delta P_t = L_t \times (\rho_s - \rho_l) \times (1 - \varepsilon_t) \times g \tag{7}$$

Where:

ΔP_t =   pressure drop across the tube due to bed weight   [N/m²]
L_t =   tube height   [m]
ρ_s =   density of the material of the solid particles [kg/m³]
ρ_l =·   density of the liquid   [kg/m³]
ε_t =   liquid volume fraction in tube or porosity   [-]
g =   earth gravity   [m/s²]

For the MSF desalination plant with stationary fluidised bed condensers specified above, the pressure drop to support the bed weight amounts to 47 000 N/m². On top of this pressure drop we have to add a pressure drop caused by the flow distribution system of 4 000 N/m² for stabilisation of the flow through all tubes. Pressure drop due to wall friction has not to be taken into account because of the very low liquid velocities in the tubes of only 0.125 m/s. However, for this particular application, we have to add the lifting height for the liquid which requires an additional pressure drop of 120 000 N/m² resulting in a total pressure drop of 47 000 + 4 000 + 120 000 = 171 000 N/m².

For the conventional MSF desalination plant we calculate a pressure drop of approx. 400 000 N/m² required by the wall friction in these very long condenser tubes with much higher liquid velocities, and when we take into account the losses in water boxes we end up with a total pressure drop of approx. 450 000 N/m².

It should be emphasised that for this particular application the pressure drop influencing the heat transfer coefficient and required by the condenser bundle installed in the conventional MSF is a factor 400 000 / 51 000 = 7.9 (!!) higher than this pressure drop for the MSF equipped with stationary fluidised bed condensers. These differences in pressure drop directly influence the pumping power requirements for both installations. In general, when also considering 'circulating' fluidised bed heat exchangers operating at somewhat higher liquid velocities and using higher density solid particles, the differences in pumping power requirements will not be that much as presented above, although, for all applications, the differences in pumping power remain easily a factor 2 to 3 times lower for the fluidised bed heat exchanger compared to the conventional shell and tube heat exchanger.

4.4 Fouling removal

Fouling of heat exchangers is experienced by a gradual and steady reduction in the value of the overall heat transfer coefficient. A closer look into this phenomenon shows that there are always two causes:

1. Fouling of the actual heat transfer surface by the forming of an insulating layer of deposits, which reduces the heat transfer through the tube wall.
2. Clogging of flow distribution system in the inlet channel and / or the inlets of the heat exchanger tubes by large pieces of dirt or deposits broken loose from the wall of vessel and piping upstream the exchanger and present in the feed flow of the exchanger. Clogging of tubes removes heat exchanger tubes from participation in the actual process of heat transfer.

The first cause can be solved by the mild scouring action of the fluidised solid particles in the tubes. The second cause, at least of the same importance as the first cause but often neglected, can only be solved by the installation of a strainer upstream the self-cleaning fluidised bed heat exchanger. To minimise the cost for such a strainer and the ground area for the heat exchanger and its accessories, we have developed a proprietary self-cleaning strainer which forms an integral part with the inlet channel of the exchanger.

Now, let us pay attention to some of our fouling removal experiences in a fluidised bed heat exchanger and, therefore, we once more should pay attention to our MSF seawater evaporators:

It is known that natural seawater cannot be heated to temperatures above 40 to 50 °C because of the formation of calcium carbonate scale. Conventional MSF evaporators often operate at maximum seawater temperatures of 100 °C, but only after chemical treatment of the seawater feed which removes the bicarbonates from the seawater and prevents the forming of scale. Of course, this is a complication in the process and does cost money. The MSF evaporator equipped with the stationary fluidised bed condensers, using 2 mm glass beads, has convincingly demonstrated that it can operate at even much higher temperatures than 100 °C without scale forming on the tube walls. Although, the scale crystals are precipitating from the seawater on the tube walls these crystals are knocked off by the glass beads at an early stage, so that it never comes to the formation of an insulating scale layer and the tube walls remain clean and shiny. Here we have clearly demonstrated the fouling removal, self-cleaning or non-fouling behaviour of a fluidised bed heat exchanger operating under harsh conditions as the result of the scouring action of the fluidised particles. No doubt that this feature is of extreme importance for heat exchangers operating on severely fouling liquids.

Meanwhile, with many self-cleaning fluidised bed heat exchangers already installed in different industries, commercial operating experiences have shown that the self-cleaning fluidised bed heat exchanger, which can remain clean indefinitely, is a cost-effective alternative to the conventional heat exchanger which suffers from severe fouling in a couple of hours, days or weeks and even months. Any type of fouling deposit, whether hard or soft; biological or chemical; fibrous, protein, or other organic types; or a combination of the above can be handled by the self-cleaning fluidised bed heat exchanger. Moreover, later in this chapter it will be shown that the unique characteristics of this heat exchange technology allow for the introduction of major design changes of installations in traditional processes and, therefore, the advantages of this heat exchange technology does reach much further than solving heat exchanger fouling problems only.

4.5 Wear

Now we have been informed about the remarkable effects of scouring particles on the heat transfer film coefficients at very low liquid velocities, low pumping power requirements and their potential to remove fouling, one might wonder what the consequences are of the scouring action of the particles with respect to wear and / or material loss of the heat exchanger tubes and the particles. After many years operating experiences we have come to the conclusion that only in case of the formation of a weak corrosion layer on tube and / or particle material, the scouring action of the particles may cause material loss due to the removal of this corrosion layer. For applications where corrosion of metal surfaces does not play a role, we present the following examples:

In a US plant, after one year of operation, the cleaning particles made of chopped stainless steel wire lost 2.5 % of weight. This is caused by the rounding-off effects of the sharp edged cylindrical particles. In the second year, being substantially rounded-off already during the first year of operation, the weight loss of the particles dropped to less than 0.5 %. Because the smooth stainless steel tube wall is not subjected to metal loss as a result of rounding-off effects, the material loss of the tubes should be much less than 0.5 % per year.

Similar experiences have been obtained in Japan with stainless steel tubes and particles. Again, after one year of operation a weight loss of approx. 2 % was measured. In the second year, this weight loss was negligible. Fig. 8 shows the rounding-off effects of chopped stainless steel wire as a function of operating time. The loss of approx. 2 % in the first year of operation as mentioned above can also be avoided by using particles which have already been rounded-off mechanically directly after their fabrication (chopping) process.

5. New installations equipped with self-cleaning fluidised bed heat exchangers

5.1 Multi-Stage Flash / Fluidised Bed Evaporator (MSF / FBE); most promising tool for thermal seawater desalination

As the development of this heat exchanger began in the early 70s with the application of stationary fluidised bed condensers in MSF evaporators, we like to begin this paragraph with what we may consider 'the origin of the fluidised bed heat exchange technology' as developed for seawater desalination, and referred to as Multi-Stage Flash / Fluidised Bed Evaporator or MSF / FBE. In the example below, we compare this evaporator with a conventional MSF. This comparison shows that the advantages of the self-cleaning fluidised bed heat exchange technology for the MSF are responsible for a much wider range of improvements than non-fouling heat exchange only.

A picture of the conventional MSF and its corresponding temperature diagram is shown in Fig. 9. The principle of this MSF can be best described as a large counter-current heat exchanger, where the cold feed is heated by the condensing vapour in the heat recovery section and the external heat supply takes place in the final heater. After leaving the final heat exchanger at its highest temperature, the liquid flashes through all stages, by way of

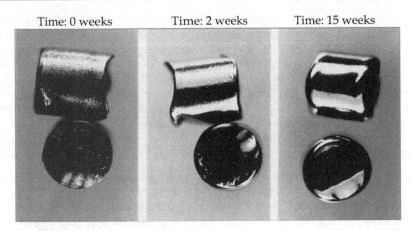

Fig. 8. Rounding-off effects of 2 mm stainless steel particles as a function of operating period.

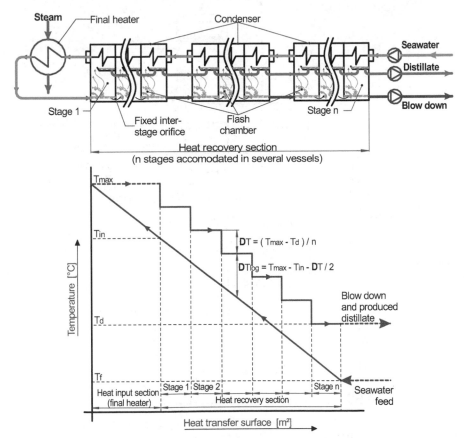

Fig. 9. Principle of conventional MSF and its temperature diagram.

openings in the bottom or intersection walls of the stages, and a gradual drop in saturation temperature takes place resulting in a partial evaporation of the liquid in each flash chamber. The flash vapour flows through the water-steam separators and finally condenses on the condenser surfaces, which are cooled by the colder incoming feed. The distillate is collected at the bottom of each stage and cascades down in the same way as the liquid in the flash chambers to the next stage. The plant has to be completed with pumps for the removal of the concentrated liquid or brine and distillate out of the coldest stage and for the feed supply. Dissolved gases and in-leaking non-condensables are removed from the feed by a vacuum line connected to a vacuum pump. The installation of the great length of horizontal condenser tubes in a conventional MSF requires the installation of several vessels in series.

The principle of the MSF / FBE is not much different from a conventional MSF, although, we have already shown that the total length of the vertical condenser tubes passing through all stages can be much shorter for an MSF / FBE than for a conventional MSF. This makes it possible to install all condenser tubes and flash chambers in only one vessel of limited height as is shown in Fig. 10.

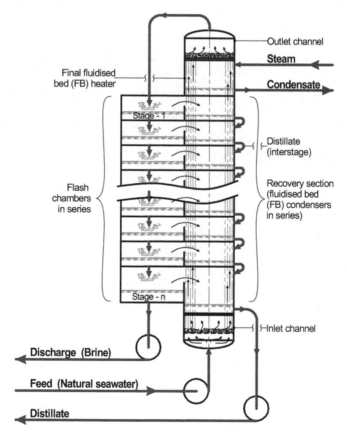

Fig. 10. Principle MSF / FBE.

As the result of the vertical layout of the MSF / FBE and the flashing down flow in the flash chambers with a height for each chamber of approx. 0.4 m, we are able to add a number of interesting improvements to the MSF / FBE in comparison with the conventional MSF, such as:

- It has already been mentioned that not only excellent heat transfer can be achieved at very low liquid velocities and very low pressure drop, but that the scouring action of the glass beads also remove scale deposits at an early stage. This makes it possible to operate the MSF / FBE at much higher maximum temperatures of the heated seawater than the conventional MSF without the need of chemicals to prevent scale. A higher maximum temperature increases the recovery of distillate from a particular seawater feed and makes it possible to design evaporators for a higher gain-ratio or lower specific heat consumption.

- *The vertical layout of the MSF/FBE* makes it possible to achieve a complete flash-off of the spraying brine flow in the flash chambers, which means that the evaporating liquid and produced vapours are in equilibrium with each other. Further, we do not need wire mesh demister for the separation of droplets from the vapour flowing into the condenser which reduces the pressure drop of the vapour flow on its way to the condenser. These advantages reduce the irreversible temperature losses in the heat transfer process, which makes it possible to save on heat transfer surface and / or reduce the specific heat consumption of the evaporator.

- *The vertical layout of the MSF/FBE* with a relatively short stage height, its high vapour space loadings, low brine levels in the flash chambers and no need for the installation of voluminous wire mesh demisters makes the flash chambers very compact, which reduces the plot area and the overall dimensions of the evaporator, and, consequently, its steel weight.

- *The vertical layout of the MSF/FBE* assures sufficient driving force for the interstage brine flow caused by the height of the brine level in the flash chambers only. Consequently, not much vapour pressure difference between stages is required to assure the interstage brine flow, which means that a large number of stages can be installed in a given flash range $\Delta T = T_{max} - T_d$ shown in Fig. 9, which again makes it possible to reduce the specific heat consumption of the evaporator.

- *The vertical layout of the MSF/FBE* makes it possible to install interstage valves between all stages using only one activator. This valve system guarantees low brine levels in all stages at varying maximum temperatures or flash range of the evaporator. A varying maximum temperature or flash range makes it possible to vary the distillate production between 0 and 100 %, while still maintaining an excellent distillate quality.

Above, we have clearly shown that *the vertical layout of the MSF/FBE*, as the result of the integration of the vertical stationary fluidised bed condenser with the flash chambers, increases the advantages of this evaporator too such an extent, that this evaporator may be considered as the most promising tool for thermal seawater desalination in the future. Fig. 11 shows an MSF / FBE demonstration plant operating on natural seawater for a distillate production of 500 m^3/d.

For more information about this fascinating evaporator, one is referred to the Ref. [8] and [9].

Fig. 11. MSF / FBE evaporator, Isle of Texel.

5.2 Reboiler at chemical plant; annual turnaround replaces cleaning every 4 to 5 days

A steam-heated evaporation system at a chemical plant in the Unites States recovers a volatile organic from a heavy organic solution laden with foulants. A hard black scale, that was forming in the upper 25 % of the tubes, was forcing the plant to switch two parallel once through rising film evaporators with a clean pair every four to five days.

When asked to increase throughput and simplify operations engineers considered installing a 190 m² falling film evaporator to operate in series with the existing rising film evaporators. Although the combination system was expected to run approximately 10 weeks between cleanings, a better solution was needed and found when engineers heard of an innovative self-cleaning fluidised bed heat exchanger technology being used at another chemical plant in the United States. The final decision in favour of the self-cleaning fluidised bed heat exchanger was made after the engineers viewed a 1 m tall, transparent desktop demonstration unit with six 12 mm up flow tubes and one 12 mm down flow (downcomer) tube.

The full-size exchanger with widened outlet channel shown in Fig. 12 at the right of the distillation column contains 73 m² of heat transfer surface. It applies internal circulation of 2.0 mm chopped stainless steel wire particles and uses 51 up flow and four down flow (downcomer) tubes according to the design shown in Fig. 4. The process liquid circulated at a constant flow of 160 m³/h, is raised from about 120 to 150 °C with condensing steam at the shell-side. Back pressure is maintained on the process side of the exchanger to prevent vaporisation which would interfere with the fluidisation of the particles. Upon discharge from the exchanger, the heated liquid flashes across a control valve into the base of the recovery column.

Comments by the operators in September 1992 after the heat exchanger had been in service for over a year without any operating problems:

There have been no shutdowns for cleaning tubes and no process upsets, and maintenance has been nil. This is a significant cost cutting result from the higher recovery of acetic acid and the more concentrated residue in the bottoms. The self-cleaning fluidised bed heat exchanger appears capable for at least a full year between turnarounds. A sample of chopped metal wire particles taken from the unit after several months of operation indicated only normal rounding off of the edges. Under the new system, the reboiler circulation rate has been constant, thus providing uniform tower operation and more total throughput. If the alternative falling film evaporator had been installed, a shutdown would have been required every 10 weeks for a costly cleaning operation.

Today, July 2011, twenty years after the heat exchanger has been put in service and after 150 000 operating hours, the heat exchanger is still in operation using the same shiny tubes and to entire satisfaction of the operators. For more information, see Ref. [2].

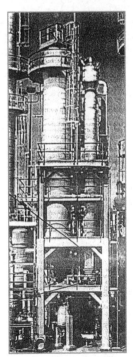

Fig. 12. Self-cleaning fluidised bed heat exchanger at chemical plant eliminates reboiler fouling.

5.3 Quench coolers at chemical plant; the real breakthrough of the self-cleaning fluidised bed heat exchange technology

A chemical plant in the United States cooled large quench water flows from a proprietary process in open cooling towers. This quench water released volatile organic compounds (VOCs) into atmosphere. As a consequence of environmental regulations the quench water cycle had to be closed by installing heat exchangers between the quench water and the cooling water from the cooling towers.

In August 1997, after considering other solutions using conventional shell and tube heat exchangers, plant management decided to carry out a test with a small self-cleaning fluidised bed heat exchanger and compared its performance with that of a conventional shell and tube heat exchanger, which suffered from a severe fouling deposit consisting of a tarry substance. Fig. 13 shows the results of this test, while Table 2 compares the design consequences for the self-cleaning heat exchangers and the conventional shell and tube exchangers. Plant management decided in favour of the self-cleaning fluidised bed heat exchange technology because of the above results and the dramatic savings on investment and operating cost.

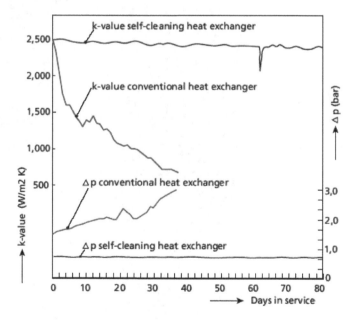

Fig. 13. Overall heat transfer coefficient (k-value) and pressure drop (Δp) as function of operating time.

	Unit	Conventional heat exchanger	Self-cleaning heat exchanger
Heat transfer surface	m²	24 000	4 600
Total number of heat exchangers	-	24 × 1 000 m²	4 × 1 150 m²
Configuration	-	3 × 50 %	2 × 50 %
Pumping power	kW	2 100	840
Number of cleanings per year	-	12	0

Table 2. Comparison conventional heat exchanger versus self-cleaning fluidised bed heat exchanger.

Fig. 14. Installation of 4 600 m² self-cleaning surface replacing 24 000 m² conventional surface.

Fig. 14 shows the installation which serves two parallel production lines. In each production line two identical self-cleaning fluidised bed heat exchangers were installed handling 2 × 700 m³/h process liquid at the tube-side and 2 × 2 100 m³/h cooling water at the shell-side. Each exchanger applies external circulation of the particles as shown in Fig. 5, has a shell diameter of 1 200 mm, a total height of 20 m and a heat transfer surface of 1 150 m², which surface consists of 700 parallel tubes with an outer diameter of 31.75 mm. Each exchanger uses 9 000 kg cut metal wire particles with a diameter of 1.6 mm.

The exchangers serving the first production line were put into operation in October 1998. Fig. 15 presents the trend of the overall heat transfer coefficient (k-value) after start-up till the end of April 1999. In spite of some fluctuations at the beginning, this figure shows a constant k-value of approximately 2 000 W/(m²·K). During a period of more than six months both exchangers operated continuously, with exception of a few short sops caused by interruptions in the power supply. The dotted line in Fig. 15 shows the trend of the k-

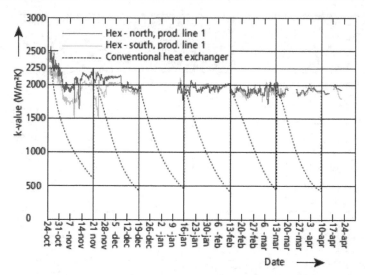

Fig. 15. k-values for self-cleaning heat exchangers of first production line, as a function of operating time and compared with the performance of conventional heat exchanger.

value for conventional shell and tube exchangers as derived from the test results shown in Fig. 13. The two exchangers of the second production line were put in operation in May 1999 and showed the same trend in k-value as the exchangers of the first production line. In December 2000, this chemical plant stopped production and the exchangers, after a final inspection, were mothballed and have never been put into operation again. This final inspection did not reveal any measurable wear of the tubes. All tubes were shiny and

	Unit	Conventional heat exchanger	Self-cleaning heat exchanger (1998)	Self-cleaning heat exchanger (2005)
Total heat transfer surface	m²	24 000	4 600	3 332
Total number of heat exchangers	-	24 × 1 000 m²	4 × 1 150 m²	4 × 833 m²
Configuration	-	3 × 50 %	2 × 50 %	2 × 50 %
Tube diameter / tube length	mm	25.4×1.65 / 12 000	31.75×1.65 / 16 000	15.88×1.21 / 8 700
Shell-side baffle type	-	segmented cross	segmented cross	EM
Total weight particles	kg	n.a.	36 000	20 000
Pumping power	kW	2 100	840	416
Number of cleanings per year	-	12	0	0

Table 3. Comparison conventional heat exchanger versus self-cleaning fluidised bed heat exchangers, state-of-the-art 1998 and 2005.

open. The cut metal wire particles showed a slight weight loss caused by rounding-off effects as discussed earlier. For more information about these fascinating heat exchange application, one is referred to Ref. [3].

In the first years of the new millennium, research and development concentrated on reducing the tube diameter of the self-cleaning fluidised bed heat exchangers in combination with rather large particles. A smaller tube diameter reduces the length of the heat exchanger tubes which creates a more compact heat exchanger with less height and, consequently, reduces the pumping power required for the process liquid. Then, we also paid attention to the installation of a novel type of baffle in the shell of the exchanger. This very innovative baffle is called the EM baffle and has been developed by Shell Global Solutions. The results of this redesign of the self-cleaning fluidised bed heat exchangers shown in Fig. 14 as far as heat transfer surface and pumping power are considered are presented in Table 3. For more information about this improved design, one is referred to Ref. [4].

6. Existing conventional severely fouling heat exchangers revamped into a self-cleaning fluidised bed configuration

The idea of changing internal circulation of particles as shown in Fig. 4 into the configuration where this circulation takes place through only one external downcomer shown in Fig. 5 was proposed by engineers of Shell in the early 90s. According to these engineers, this modification would make it possible to revamp existing vertical severely fouling conventionally designed reboilers into a self-cleaning configuration. Moreover, it would be an elegant and rather low cost but also a low risk approach to introduce a new technology due to the possibility of an immediate fallback from new technology to old proven technology. This idea is not only applicable for reboilers but also for evaporators and crystallisers and the constant circulating flow required by these unit operations corresponds with the preferred type of flow for the self-cleaning fluidised bed heat exchanger.

Moreover, also in this paragraph, it will be shown that this approach of introducing the self-cleaning heat exchange technology in existing plants could not only be an attractive solution for straight forward rather simple heat exchange applications suffering from severe fouling, but also for very complex industrial processes.

6.1 Reboiler

A typical example of a conventional reboiler that is suitable for revamping is shown in Fig. 16 with the revamped configuration shown in Fig. 17. Generally speaking, the requirements specified by plant management for the majority of revamps can be summarised as follows:

1. The same process conditions should be maintained as in the original installation, i.e. flow, temperatures and liquid velocity in the tubes.
2. The connections to the column should be maintained.
3. The installed pumps should be used and can be used because pumping power requirements are generally lower for the self-cleaning fluidised bed heat exchangers than for the conventional severe fouling shell and tube exchanger.

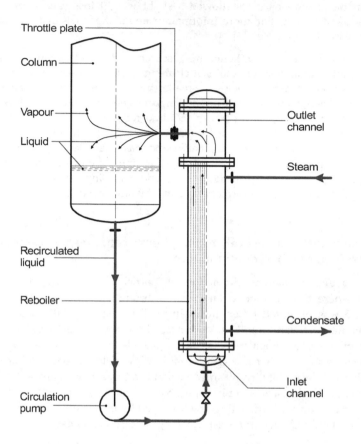

Fig. 16. Evaporator equipped with conventional heat exchanger.

4.   As many components of the existing installation should be used in the revamped configuration like bundle, channels or maybe even modified channels.
5.   The revamp must be carried out within the available space. This often means that a revamp can only be carried out when the existing installation has already a vertical position.

The advantages of most revamps are much lower maintenance cost, an increased production and 'smoother' operation.

6.2 Cooling crystallisation plant

A 2-stage cooling crystallisation plant in Egypt produces Sodium Sulphate. The chillers of both stages suffer from severe fouling caused by heavy deposits of crystals. Shutdown of the installation every 24 hours for melting out these deposits is common. The conventional cooling crystalliser is shown in Fig. 18, while Fig. 19 depicts this installation after its revamp into a self-cleaning configuration.

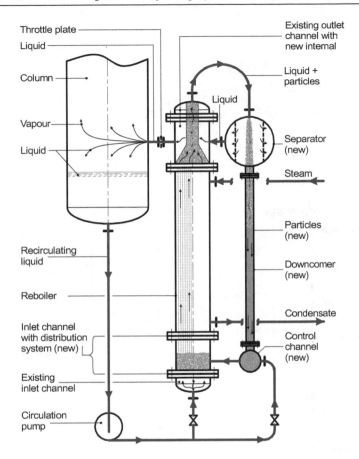

Fig. 17. Conventional evaporator revamped into a self-cleaning configuration.

Calculations have shown that the investments necessary for the modification of the existing installation into a self-cleaning configuration will be paid back by a substantially increased production in approximately six months.

Information about the reboiler, the cooling crystallisation plant and other applications discussed in this paragraph can be found in Ref. [7]

6.3 Evaporator for concentration of very viscous severely fouling slurry

In one of the Scandinavian countries, a production plant of a proprietary product operates a very large MVR evaporator for the concentration of a slurry up to approx. 70 % solids. Even at a temperature of 100 °C this slurry, which behaves non-Newtonian, has a very high viscosity varying between 50 and more than 200 cP. This very large shell and tube heat exchanger suffers from severe fouling which sometimes requires one month (!!) of mechanical cleaning after only three months (!!) of operation. In Fig. 20, the test plant in parallel with the existing evaporator is shown and the dimensions of the existing installation

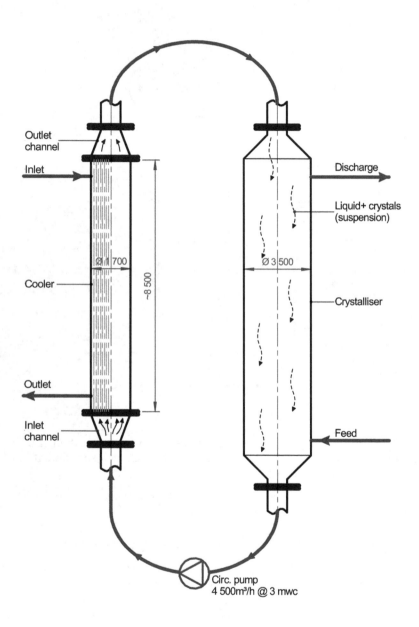

Fig. 18. Conventional cooling crystalliser.

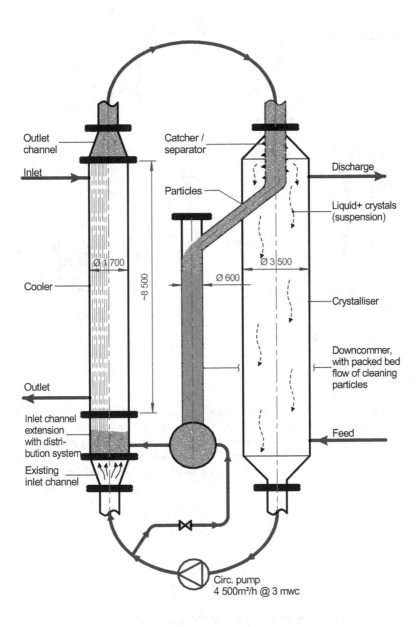

Fig. 19. Revamped into self-cleaning configuration of Fig. 18.

give a good impression about its size, although provided with relatively small diameter tubes with an ID of only 20 mm.

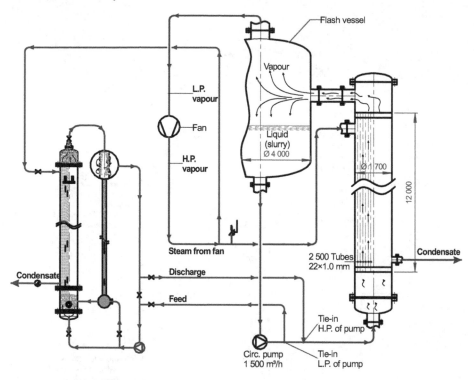

Fig. 20. Existing evaporator and test installation.

The proposal for the revamp of this installation is shown in Fig. 21 and uses a maximum of the very large existing components, including the circulation pump. The first series of experiments with the test installation are promising and shear-thinning effects caused by the increased turbulence of the slurry induced by the action of the fluidised particles are reducing the viscosity of the slurry substantially and have produced heat transfer coefficients or k-values between 1 000 and 2 000 W/(m² K) depending on the concentration of the slurry without fouling. These coefficients should be compared with the clean heat transfer coefficients of approximately 600 W/(m² K) for the conventional heat exchanger which, in a couple of months, reduces to only a fraction of its clean value due to by fouling.

This potential revamp reflects the benefits of recent developments which make it possible to operate a self-cleaning fluidised bed heat exchanger on a very viscous slurry and use rather large stainless steel particles (2.5 mm) in small tube diameter with an ID of only 20 mm.

6.4 Combination of preheater and thermal siphon reboiler

A chemical plant in the United States operates the preheater in series with the thermal syphon reboiler shown in Fig. 22. The 8-pass preheater with tubes with an O.D. of 25 mm, a very high liquid velocity in the tubes of 4.5 m/s and heated by L.P. steam experiences severe fouling still requiring cleanings every two months, while the thermal syphon reboiler heated by M.P. steam requires cleanings every four months.

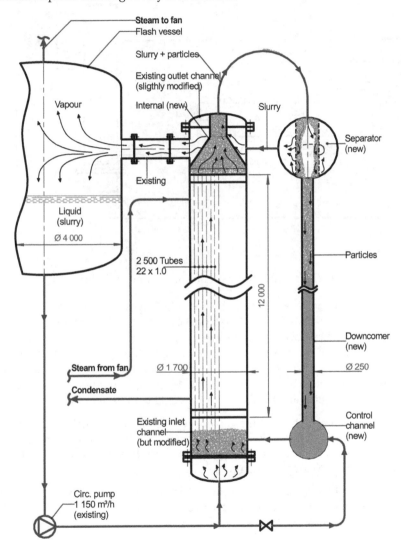

Fig. 21. Existing evaporator revamped into self-cleaning configuration.

The solution we are proposing to solve this problem is quite unique and explained in Fig. 23. As a matter of fact, we have increased the tendency of fouling in the preheater due to the precipitation of solids by increasing the outlet temperature of the preheater. This can be realised by adding M.P. steam to the shell of the preheater instead of L.P. steam. As a result of this temperature increase, the preheater will also partly contribute to the degassing

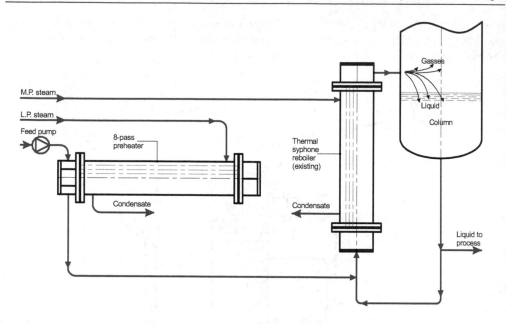

Fig. 22. Conventional preheater in series with thermal syphon reboiler.

of the liquid which is normally done in the reboiler. Above goals have been realised by revamping the existing 8-pass horizontal conventional heat exchanger into a vertical single-pass self-cleaning fluidised bed configuration using stainless steel cleaning particles with a diameter of 2.5 mm and also installing an extra circulation pump to maintain sufficient velocity in the tubes of our single-pass configuration for circulation of the cleaning particles. Although, we have indeed increased the tendency for fouling, we expect that the introduction of our self-cleaning technology will keep the preheater clean.

The separation of the gasses from the mixture of liquid and particles takes place in the widened outlet channel of the preheater, these gasses are fed into the reboiler and evenly distributed over all the tubes of the reboiler where they contribute to the (natural) circulation effect of this thermal syphon reboiler.

Considering the fact that a substantial fraction of the totally required degassing is not done anymore in the reboiler, the heat load of the reboiler can be reduced, which reduces the condensing steam temperature, the tube wall temperature and, consequently, the fouling of the reboiler.

The advantage of this approach is the revamp of the conventional preheater into a self-cleaning configuration at an increased heat load. An experiment with a single-tube self-cleaning pilot plant in parallel with the existing severely fouling preheater should demonstrate the non-fouling performance of the self-cleaning heat exchange technology. If this is indeed the case, then, we have not only solved the fouling problem of the preheater at an even higher heat load, but also reduced the fouling of the conventional thermal syphon reboiler.

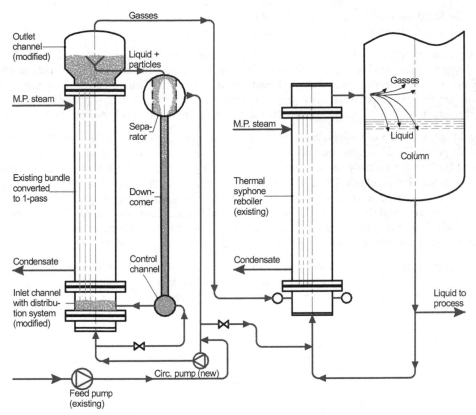

Fig. 23. Conventional preheater revamped into self-cleaning configuration and operating in series with thermal syphon reboiler.

For the proposed solution of this problem, we have introduced the concept of evaporation of a fraction of the liquid creating a mixture of liquid, vapour and particles in the tubes. We know that this is possible if certain design criteria are taken into account. Consequently, with this example, we have presented the possibility that our self-cleaning heat exchange technology can also be applied for applications where we even experience boiling or evaporation in the tubes.

6.5 Self-cleaning fluidised bed heat exchangers in existing 'directly heated' HPAL plants

There exist a strong drive to apply indirect heating (i.e. using heat exchangers) in High Pressure Acid Leach (HPAL) plants for the extraction of nickel and cobalt from laterite ore slurry, because of the benefits of indirect heating in comparison with direct heating (i.e. using steam injection or slurry / vapour mixing condensation), which benefits we summarise below:

- Increased autoclave production capacity.
- Reduced acid consumption.

- Reduced neutralizing agent consumption.
- Recovery of demineralised condensate and process condensate.

Poor heat transfer and hydraulic performance of conventional shell and tube heat exchangers have worked against the introduction of indirect heating in HPAL plants. We believe that self-cleaning fluidised bed heat exchangers offer a much better option, and in the example below, we introduce a 'directly heated' HPAL plant which is retrofitted into an 'indirectly heated' configuration using two different kinds of heat exchangers.

Fig. 24 shows the flow diagram, including relevant temperatures, of a 'directly heated' HPAL plant. Fig. 25 shows the above flow diagram, but, now extended in such a way that direct heating can be fully replaced by indirect heating. Now, for the high temperature end of the installation shown in Fig. 25, we have engineered these two different kinds of indirect heating solutions. One of the indirect heating solutions uses conventional shell and tube heat exchangers and the other self-cleaning fluidised bed heat exchangers. Table 4 compares both indirect heating solutions. The advantages in favour of the self-cleaning fluidised bed configuration are very convincing and we like to emphasise these advantages:

- Shear-thinning of the non-Newtonian highly viscous slurry due to the increased turbulence of the slurry induced by the fluidised particles, which reduces the viscosity of the slurry experienced by the fluidised bed by a factor 4 to 5 or even more.
- High heat transfer coefficients,
- low slurry velocities,
- low pressure drops in the tubes, and
- non-fouling due to the scouring action of the fluidised particles on the tube wall.

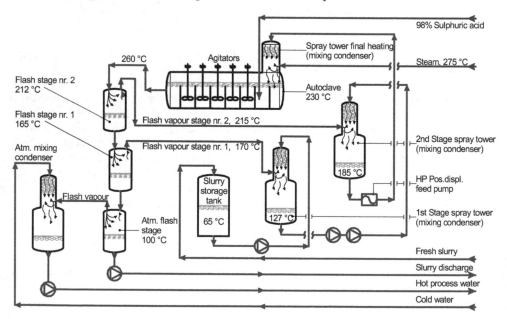

Fig. 24. HPAL plant for laterite nickel employing direct heat transfer.

Particularly, the high heat transfer coefficient and low slurry velocity do affect the total length of the installed heat exchange tubes. This follows from the Equation (5) for the tube length presented in paragraph 0 of his chapter, after substitution of the design and process parameters. As a consequence, the number of shells in series for the self-cleaning fluidised bed heat exchanger is a fraction (just one) in comparison with the large number of shells in series for the conventional shell and tube heat exchanger.

For this HPAL application, the scope of the benefits already mentioned at the beginning of this sub-paragraph increases when indirect heating is not only applied to the highest temperature stage of the installation but to all stages. It is not surprising that all major mining companies show much interest in the self-cleaning fluidised bed heat exchange technology for an even greater variety of applications than only HPAL for the extraction of metals from laterites.

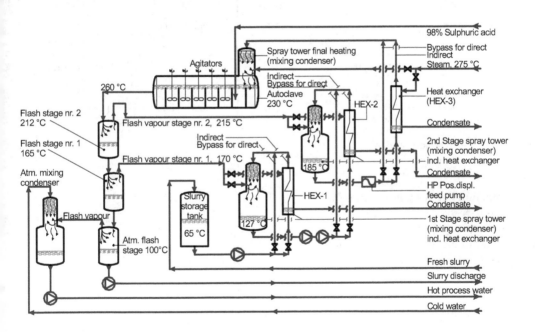

Fig. 25. HPAL plant for laterite nickel employing direct heat transfer revamped into indirect heated configuration.

For more information about the performance and the potential of HPAL plants equipped with self-cleaning fluidised bed heat exchangers, one is referred to Ref. [6].

	Unit	Conventional shell and tube	Self-cleaning fluidised bed
Inlet- / Outlet- / Steam temperature	°C	185 / 235 / 275	185 / 235 275
Density slurry	kg/m³	1 340	1 340
Specific heat slurry [kJ/(kg K)]	kJ/(kg K)	3.6	3.6
Dynamic viscosity	cP	50 - 70	10 - 15
Tube velocity slurry	m/s	2.0	0.35
Diameter tube	mm	38 × 3.0	38 × 3.0
Diameter- / Material particles	mm	n.a.	4.0 / Titanium
Clean- / Design k-value	W/(m²·K)	~ 600 / 300	1 500 / 1 500
Tube length based on design k-values and Eq. (5)	m	166.8	5.84
Total number of shells in series for 1-pass tube-side and tube length per shell equal to 8 m	-	21	1
Total number of shells in series for 2-pass tube-side and tube length per shell equal to 8 m	-	11	n.a.
Pressure drop	bar	~ 6 - 10	< 1.0

Table 4. Comparison significant parameters for indirect heating of high temperature stage of HPAL plant of Fig. 25.

7. Final remarks

We have given an indication about the cost of fouling of heat exchangers on a global scale and we have shown that the self-cleaning fluidised bed heat exchange technology can play a significant role in battling these fouling cost, and does have even more potential that solving fouling problems only.

Particularly, the latter aspect has caught the attention of an increasing number of very large companies which are very much interested to implement the self-cleaning fluidised heat exchange technology for the upgrading of their existing proprietary processes, or even for the development of completely new processes.

8. References

Garrett-Price, B.A., et al. (1985). *Fouling of Heat Exchangers*, Noyes Publications, Parkridge, New Yersey.

Gibbs, R. & Stadig, W. (1992). *Fluidized bed heat exchanger eliminates reboiler fouling*, Chemical Processing, August.

Klaren, D.G. (2000). *Self-Cleaning Heat Exchangers: Principles, Industrial Applications and Operating Installations*, Industrial Heat Transfer Conference, Dubai, UAE, September.

Klaren, D.G. & de Boer, E.F. (2004). *Case Study Involving Severely Fouling Heat Transfer: Design and Operating Experience of a Self-Cleaning Fluidized Bed Heat Exchanger and its Comparison with the Newly Developed Compact Self-Cleaning Fluidized Bed Heat Exchanger with EM Baffles*, Presented at the Fachveranstaltung: Verminderung der Ablagerungsbildung an Wärmeübertragerflächen, Bad Dürkheim, Germany, October.

Klaren, D.G., de Boer, E.F. & Sullivan, D.W. (2007). *Consider low fouling technology for 'dirty' heat transfer services*, Hydrocarbon Processing, Bonus Report, March.

Klaren, D.G., de Boer, E.F. & Crossley, B. (2008). *Reflections on Indirect Heating of Laterite Ore Slurry in HPAL Plants Using Self-Cleaning Fluidised Bed Heat Exchangers*, Presented at ALTA 2008, Perth, Western Australia, June.

Klaren, D.G. & de Boer, E.F. (2009). *Revamping existing severely fouling conventional heat exchangers into a self-cleaning (fluidised bed) configuration: New developments and examples of revamps*, International Conference on: Heat Exchangers Fouling and Cleaning-2009, Schladming, Austria.

Klaren, D.G. and de Boer, E.F. (2010). *Multi-Stage Flash / Fluidized Bed Evaporator (MSF / FBE): A resurrection in Thermal Seawater Desalination?*, CaribDA Conference, Grand Cayman, June.

Klaren, D. G. (2010). *Design, Construction and Operating Features of Multi-Stage Flash / Fluidized Bed Evaporators (MSF/FBE) for very large Capacities*, IDA Conference, Huntington Beach, California, USA, November.

Richardson, J.F. and Zaki, W.N. (1954). *Sedimentation and fluidization: Part 1*, Trans. Inst. Chem. Eng., vol. 28, p. 35, 1954.

Ruckenstein, E. (1959). *On heat transfer between a liquid/fluidized bed and the container wall*, Rev. Roum. Phys., vol 10, pp. 235-246.

6

Fouling and Fouling Mitigation on Heat Exchanger Surfaces

S. N. Kazi

Department of Mechanical and Materials Engineering,
Faculty of Engineering,
University of Malaya, Kuala Lumpur,
Malaysia

1. Introduction

Heating or cooling of one medium by another medium is performed in a heat exchanger along with heat dissipation from surfaces of the equipment. In course of time during operation, the equipment receives deposition (Fouling) which retards heat exchanging capability of the equipment along with enhanced pressure loss and extended pumping power. Thus accumulation of undesired substances on a surface is defined as fouling. Occurrence of fouling is observed in natural as well as synthetic systems. In the present context undesired deposits on the heat exchanger surfaces are referred to fouling. With the development of fouling the heat exchanger may deteriorate to the extent that it must be withdrawn from service for cleaning or replacement.

The overall design of heat exchanger may significantly be influenced by fouling, use of material, process parameters, and continuous service in the system or process stream are all deliberately influenced by fouling phenomena. Preventive measures of fouling are highly encouraged as it keeps the service of heat exchanger for a longer time. However many mitigation techniques of fouling are harsh to the environment. A technique involving chemicals and means benign to the environment is the most desired approach and it could elongate the cleaning interval. On the other hand unique and effective arrangements may be required to facilitate satisfactory performances between cleaning schedules. As a result fouling causes huge economic loss due to its impact on initial cost on heat exchanging operation, operating cost, mitigation measures and performance. The present study focused on fouling phenomena, fouling models, environment of fouling, consideration of heat exchanger fouling in design and mitigation of fouling.

2. Fouling

Fouling is the resultant effect of deposition and removal of deposits on a heat exchanger surface. The process of fouling could be represented by the equation (2.1).

$$\frac{dm_f}{dt} = \dot{m}_d - \dot{m}_r \tag{2.1}$$

where dm_f, $\dot{m}_d$ and $\dot{m}_r$ are net deposition rates, deposition and removal rates respectively.

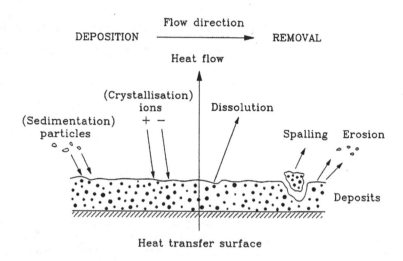

Fig. 2.1. Various deposition and removal processes during fouling.

Various deposition and removal processes for a typical system could be predicted as shown in Figure 2.1. The processes occur simultaneously and depend on the operating conditions. Usually removal rates increase with increasing amounts of deposit whereas deposition rates are independent of the amount of deposit but do depend on the changes caused by deposits such as increase in flow velocity and surface roughness. In the application of constant wall temperature or constant heat transfer coefficient boundary conditions, the interface temperature decreases as deposits build up which reduces the deposition rate.

Initiation period or time delay in heat exchanger fouling is considered the time when there is no deposition for some time after a clean heat exchanger has been brought into operation. Figure 2.2 illustrates this in detail. The initial growth of deposit can cause the heat transfer coefficient to increase rather than decrease resulting in a fouling resistance due to changing flow characteristics near the wall. At the initial stage the deposit penetrates the viscous sub-layer, the resulting turbulence increases the film heat transfer coefficient at the solid/liquid interface by changing flow characteristics near the wall. This increase in heat transfer coefficient may overcome the thermal resistance offered by the deposits and the net heat transfer coefficient may increase.

Several authors have reported negative fouling resistances [1, 2]. This process continues until the additional heat transfer resistance overcomes the advantage of increased turbulence. The time period from the beginning of the fouling process until the fouling resistance again becomes zero is called roughness delay time [3]. The time period from the beginning, when the formation of stable crystalline nuclei and their concretion to a compact fouling layer takes place is also called as induction period, which is in fact the roughness delay time and it ends up with the increase of fouling resistance above zero level.

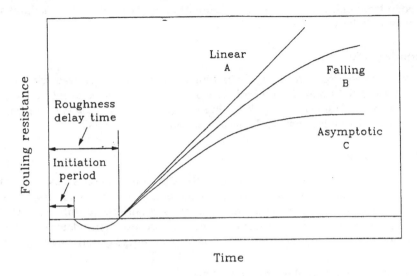

Fig. 2.2. Typical fouling curves.

The initiation period and the roughness delay time for particulate fouling are very small [4] in comparison to the fairly long delay time for crystallization fouling [5]. After the roughness delay time, the fouling curve can be classified into three categories, (a) Linear, (b) Falling, and asymptotic, as illustrated in Figure 2.2.

The linear fouling curve is obtained for very strong deposits where removal is negligible or in case where the removal rate is constant (and deposition is faster than removal). The falling rate curve is obtained from decrease in deposition and deposits with lower mechanical strength. The combined effect with time causes the net deposition or fouling rate to fall. Asymptotic fouling curve has been most commonly reported for different types of fouling. The removal rate increases with time for weak deposits and can eventually become equal to the deposition rate. The net rate is then zero as depicted in Figure 2.2.

Linear fouling curves have been presented by many authors for crystallization fouling [6-8]. However, there is some doubt as to whether the fouling rate may remain linear for a long time. For the constant heat flux situation, the net driving force may decrease with fouling. The increase in flow velocity due to the reduced cross-sectional area with deposit formation can increase the removal rate and the linear rate may change to a falling rate or even level-off completely [9]. Asymptotic behavior for crystallization fouling has reported by various authors [10, 5, 11-12]. Cooper et al. [13] found asymptotic behavior for calcium phosphate fouling (with some particulate fouling from suspended solids). For particulate fouling, asymptotic behavior is attained because particles do not adhere strongly to the wall and can be removed easily [4, 14].

A fouling process that follows a linear rate for constant heat flux can have falling or even asymptotic behaviour for constant temperature difference. The interface temperature decreases with deposit formation because of the extra resistance offered by deposit layer and enhanced flow velocities as flow passages are partially blocked by deposits. Thus the thermal boundary conditions can result in different fouling curves which may give wrong impressions about the actual fouling mechanism.

2.1 Categories of fouling

Fouling can be categorised a number of different ways. These are (1) heat transfer service, (2) type of service fluid and (3) application. Most fouling situations are virtually unique. Fouling [15] can be classified into the following categories: (i) particulate, (ii) Precipitation, (iii) corrosion, (iv) biofouling and (v) chemical reaction.

2.2 Particulate fouling

Particulate fouling is evolved by the accumulation of solid particles suspended in the process stream onto the heat transfer surface. Heavy particles settle on a horizontal surface due to gravity and fine particles settle onto heat transfer surfaces at different inclinations due to suction force or other mechanisms. Unburned fuels or ashes deposition on boiler tubes, dust deposition on air cooled condensers etc. are examples of particulate fouling.

2.3 Precipitation fouling (sedimentation fouling)

This kind of fouling is also called crystallization fouling. Dissolved inorganic salts are normally present in fluid used in heat exchanger. There is a maximum amount of the salt (saturated) which can be dissolved in this fluid. During heating or cooling supersaturation occurs in the dissolved inorganic salts. The inverse solubility salts such as calcium and magnesium sulphate, carbonate, silicate, etc. have less solubility in warm water up to a certain temperature than in cold water. This may occur when the process condition inside the heat exchanger is different from condition at the entrance. A stream on a wall at a temperature above that of corresponding saturation temperature for the dissolved salts allows crystal formation on the surface. Normally crystallization starts at especially active points – nucleation sites – such as scratches and pits and often after induction period spread to cover the entire surface. This type of fouling is strong and adherent and requires vigorous mechanical or chemical treatment to be removed [16]. Fouling rate increases with the increase of salt concentration or surface temperature. These are often found in heat exchangers of process industries, boilers, evaporators etc.

2.4 Chemical reaction fouling

This type of fouling occurs when the depositions are formed as a result of chemical reaction resulting to produce a solid phase at or near the surface. In the present case carbonaceous material deposits due to thermal gradation of the components of a process stream on hot heat transfer surface. This type of fouling is often extremely tenacious and need special measure to clean off the deposits from heat exchanger surfaces to provide them satisfactory operation life [16].

2.5 Corrosion fouling

This type of fouling is also caused by some chemical reaction but it is different from chemical reaction fouling. This fouling is a reactant and it is consumed. In this case, the surface reacts with the fluid and become corroded [15]. The corrosion products can foul the surface provided it is not dissolved in the solution after formation. pH value of the solution is one of the controlling parameter. Such as, presence of sulfur in fuel can cause corrosion in gas and oil fired boilers. Corrosion is often more prone in the liquid side of the heat exchanger. In some cases the product of corrosion may be swept away to downstream of a process loop and cause deposition on surfaces there.

2.6 Accumulation of biological fouling

On a heat transfer surface the growth of biological materials results in biofouling. In this case biological micro and macro organisms are stick to the heat transfer surface. When microorganisms (e.g., algae, bacteria, molds etc.) and their products grow they form microbial fouling. Seaweeds, waterweeds, barnacles develop microbial fouling. These fouling may occur simultaneously. The growth of attached organisms is one of the common problems [15] in heat exchanger operation. Food processing industries, power plant condensers using seawater, etc. are experiencing biofouling.

2.7 Fouling process

Fouling is a complex phenomenon due to involvement of a large number of variables. From a fundamental point of view the fouling mechanism follows certain stages in developing on a surface [17]. These are: Initiation, transport, attachment, removal and aging.

2.8 Initiation

Surface is conditioned in the initiation period. The initial delay induction period is influenced by the materials surface temperature, material, surface finish, roughness and surface coating. With the increase of degree of supersaturation with respect to the heat transfer surface temperature or increase of surface temperature the induction period decreases. During the induction period, nuclei for crystallization of deposit are also formed for biological growth. This period can take a long time, may be several weeks or a few minutes or even seconds.

The delay period decreases with increasing temperature in chemical reaction fouling due to the acceleration of induction reactions. If the initial period decreases with increasing surface temperature, crystallization fouling would be changed [18]. With the increase of surface roughness the delay period tends to decrease [19]. Additional sites are developed by the roughness projections, which promotes crystallization while grooves provide regions for particulate deposition.

2.9 Transport

In this part, fouling substances from the bulk fluid are transported to the heat transfer surface across the boundary layer. This is dependent on the physical properties of the system and concentration difference between the bulk and the surface fluid interface. Transport is accomplished by a number of phenomena including diffusion, sedimentation and thermophoresis [20, 21]. The local deposition flux $\dot{m}_d$ on a surface can be expressed by equation (2.1).

$$\dot{m}_d = h_D\left(C_b - C_s\right)$$

(2.1)

Where, C_b and C_s are reactant concentration in the bulk fluid and that in the fluid adjacent to the heat transfer surface where as h_D is the convective mass transfer coefficient. From Sherwood number ($Sh = h_D d / D$), h_D could be evaluated. Sherwood number is dependent on the flow and the geometric parameters.

The phenomenon of transportation of a particulate matter in a fluid due to gravity on a horizontal or inclined surface is known as sedimentation. This is playing a vital roll where particles are heavy and fluid velocities are low.

2.10 Attachment

At this stage, the deposits are adhered to the surface and among itself. Salt ions approaching to the surface are attracted to it due to electro-magnetic forces and adhere to the surface to form nucleation and gradually it grows with time to form a fouling layer. Thus forces acting on the particles as they approach the surface are impotent in determining attachment. Properties of the materials, such as size, density and surface conditions are dominating the attachment phenomenon.

2.11 Removal

There is competition between removal and deposition of the foulants, up to the steady growth of the deposition on the surface. Shear forces at the interface between the fluid and deposited fouling are responsible for removal. The velocity gradients at the surface, the viscosity of the fluid and surface roughness are guiding the shear forces. Removal from the surface is performed through the mechanism of dissolution, erosion and spalling.

2.12 Aging

With the commencement of deposition ageing starts. During the ageing, there may be transformation of crystal to improve or decrease the deposition strength with time. During aging the mechanical properties of the deposit can change due to changes in the crystal or chemical structure. Alteration of the chemical composition of the deposit by a chemical reaction may change its mechanical strength. On the other hand the biofouling layer may become weak due to corrosion at the surface by slow poisoning of microorganisms.

2.13 Change in deposition thickness with time

Figure 2.3 is showing the growth rate of deposit on the surface [15]. Region A: fouling is initiated in the induction period. Region B: a steady deposit growth on the surface. The rate of removal of deposit was increased when the rates of deposition gradually retards. Region C: in this region the rate of removal and deposition seems equal and the thickness of deposition remains constant.

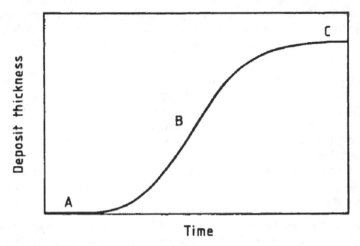

Fig. 2.3. Change in deposition thickness with time.

2.14 Composite fouling

Some of the common salts causes fouling are $CaSO_4$, $CaCO_3$ and $Mg(OH)_2$, and SiO_2. Solubility, crystal structure and strength have impact on composite scale formation in fouling. Therefore, composite fouling needs more attention and further research [17].

3. Effects of fouling

Fouling phenomena imposes retardation on heat transfer and augmentation of frictional pressure drop which degrades the effectiveness of a heat exchanger. Some basic design aspects of heat exchangers along with mitigation of fouling are discussed in the present chapter.

3.1 Effect of fouling on heat exchanger design

A fixed value of fouling resistance could be assigned during the design stage although fouling is time dependent phenomenon. The cleaning schedule and operating parameters of the heat exchanger is dependent on the design fouling factor. Depending on application some heat exchangers require frequent cleaning whereas some need rear cleaning. Fouling rate is a dominating factor in designing a particular heat exchanger.

Fouling allowance: Provisions are during the design stage once fouling is anticipated. Different approaches are used to provide an allowance for fouling resistance. They all result into an excess heat transfer surface area. Updated methods include, specifying the fouling resistances, the cleanliness factor, or the percentage over surface.

A fouling resistance is prescribed on each side of the surface where fouling is anticipated. A lower overall heat transfer coefficient is resulted. To achieve the specified heat transfer, excess surface area is provided. Until the specified value of the fouling resistance is reached, the performance of the heat exchanger will be satisfactory. Depending on this fact, maintenance schedule could be planned to avoid unprecedented shut down for cleaning.

Tubular Exchanger Manufacturers Association (TEMA) [22] is referenced source of fouling factors used in the design of heat exchangers. Plant data, proprietary research data, personal and company experience etc. are other sources of fouling resistance data could be used in design.

Minimize Fouling by considering Design Features: Extent of fouling could be minimized by good design practice. Direct contact heat exchangers are considered where excess fouling is desired. In general a fouling prone fluid stream should be placed on the tube side as cleaning is easier. Generally higher fluid velocity and lower tube wall temperature retard fouling accumulation. Velocity of 1.8 m/s is a widely accepted figure for tube side flow of a heat exchanger. Heat Exchangers, operating over dew point for acid vapor and above freezing for fluids containing waxes prevent corrosion and freezing fouling. Fouling deposits are always found heavy in the region of low velocity at the vicinity of baffles in the shell side of the shell and tube heat exchangers.

Design features to facilitate fouling control: Full elimination of fouling may not be possible by good design practice alone. So, heat exchangers require cleaning at certain intervals. On-line cleaning can be employed to control fouling by extending cleaning cycle. Continuous fouling can ensure minimized fouling allowance. At construction and installation phase of a plant on-line cleaning system could be installed at ease. A heat exchanger with removable head and straight tube would be easy to clean and maintain. Space and provision for removal and cleaning of tube bundles are required to be available. On site cleaning facilities are to be provided with options of keeping isolation valves and connection provisions for cleaning hoses which could lead to chemical cleaning.

Fouling and operation of heat exchangers: Provision of excess surface area in heat exchangers for curbing fouling may lead to operation problem and fouling build. Generally high heat transfer area enhances total heat transfer which raises the out let temperature. By changing process parameters such as flow, surface temperature leads to higher fouling.

Fouling control strategies: A number of strategies are applied for fouling control. In operating condition additives are added. On-line or off-line surface cleaning techniques are other options. To control fouling under different consequences are consolidated by some researchers as stated in Table 3.1 [23].

On-line techniques	Off-line techniques
Use and control of appropriate additives: Inhibitors, Antiscalants, Dispursants, Acids, Air jet	Disassembly and manual cleaning: Lances: Liquid jet, Steam, Air jet. Mechanical Cleaning:
On-line cleaning: Sponge balls, Brushes, Sonic horns, Soot blowers, Chains and scrappers, Thermal shock, Air bumping	Drills, Scrapers Chemical cleaning

Table 3.1. Various techniques adapted to control fouling.

Heat Exchanger with green additives: Many additives were developed for retardation of fouling but many of them found carcinogenic in nature. Now researchers are heading towards green additives. Chemistry and analysis are underway. Lab analysis and performances will be subsequently achieved. In near future users are looking for a breakthrough in this field.

3.2 Fouling effect on heat transport

Mineral scales deposited on heat exchanger surfaces are a persistent and an expensive problem in process industries, cooling water systems, steam generation units, desalination by evaporation etc. and also house hold equipment. Precipitation of mineral salts as a scale on the surface of the conduit and cause obstruction of fluid flow, impedance of heat transfer, wear of metal parts, localized corrosion attack and unscheduled equipment shutdown.

The deposit layer provides an additional resistance to heat transfer. Generally, the thermal conductivity of the deposit layer is very low compared with that of the material of the heat exchanger which may result in a much higher thermal resistance than the wall or film resistances. The deposit layer also reduces the flow area, which increases the pressure drop. This problem is quite severe and is further enhanced by the rough surface of the deposit. Both effects reduce the heat exchanger performance significantly. Additional energy requirements in terms of more heating or pumping power can hamper the economics of the process.

In a circular tube, fouling builds on the inside or outside of the tube depending on the flowing fluid. Fouling adds an insulating cover to the heat transfer surface. The overall heat transfer coefficient for a smooth tubular heat exchanger under deposited conditions, U_f can be obtained by adding the inside and outside thermal resistances:

$$U_f = \frac{1}{A_o / A_i h_i + A_o R_{fi} / A_i + A_o \ln(d_o / d_i) / 2\pi kL + R_{fo} + 1 / h_o} \tag{3.1}$$

where R_{fi} and R_{fo} represent resistances for the outer and inner surfaces of the tubes.

The thermal resistance due to fouling is evaluated generally based on experiments as difference in the overall specific resistances of the fouled and clean wall:

$$R_f = \left(\frac{1}{U_f} - \frac{1}{U_{cl}} \right) \tag{3.2}$$

Where, the overall heat transfer coefficient U_f can also be evaluated by using the rate equation:

$$U_f = \frac{\dot{Q}}{\left(A \times \Delta T_f \right)} \tag{3.3}$$

The heat flow rate $\dot{Q}_f$ and temperature difference ΔT_f (the temperature difference between heated surface and the bulk liquid) are experimentally obtained. A is the exposed area of the heat exchanging surface to the liquid. The net rate of deposition of $CaSO_4.2H_2O$ on metal surface is estimated as $\frac{m}{t}$, where m is the total mass accumulation on a unit area and t refers to the amount of time the surface was exposed to the solution of the foulant.

Using the definition of heat transfer coefficient and fouling resistance, the equation (3.4) can be derived for constant heat duty.

$$\frac{A_{fouled}}{A_{clean}} = 1 + U_{clean}R_f \tag{3.4}$$

The required excess heat transfer area usually becomes excessive due to the higher clean heat transfer coefficients. It is often recommended that the additional surface should not exceed 25 percent of the heat transfer surface requirement for clean operation.

3.3 Effect of fouling on pressure drop

In heat exchangers pressure loss is considered more critical than loss in heat transfer due to fouling. Fouling results in a finite layer. Flow field, pressure drop are affected by the change in geometry of the flow passage. Thus in a tubular heat exchanger, the deposited layer roughens the surface, diminishes the inner and raises the outer dimension of the tubes. The inside diameter of the tube decreases and roughness of the tube increases due to fouling which, causes an increase in pressure drop. Pressure drop inside a tube of a heat exchanger under fouled and clean state can be correlated as follows:

$$\frac{\Delta P_f}{\Delta P_c} = \frac{f_f}{f_c}\left(\frac{d_c}{d_f}\right)\left(\frac{u_{mf}}{u_{mc}}\right)^2 \tag{3.5}$$

Considering that the mass flow rates under clean and fouled conditions are the same, the mass flow rate can be represented as:

$$\dot{m} = \rho u_m A_{cr} \tag{3.6}$$

Equation (3.5) thus becomes:

$$\frac{\Delta P_f}{\Delta P_c} = \frac{f_f}{f_c}\left(\frac{d_c}{d_f}\right)^5 \tag{3.7}$$

The magnitude of d_f of scaled tube can be obtained from equation (3.8).

$$d_f = d_c \exp\left(-\frac{2k_c R_f}{d_c}\right) \tag{3.8}$$

The thickness t_f of deposit layer can be obtained from:

$$t_f = 0.5d_c \left[1 - \exp\left(-\frac{2k_f R_f}{d_c} \right) \right] \tag{3.9}$$

For a known total fouling resistance, the tube diameter under fouled conditions can be evaluated on knowing the thermal conductivity of the deposits. Non-uniform thermal conductivity may result from the multi layers of fouling deposits. Approximate thermal conductivities of pure materials constituting fouling deposits are often used for estimation of thermal conductivity of the total deposits. Depending on situations the fouling layer is considered composed solely of one material. In some occasions to ease calculations f_f is considered equal to f_c.

4. Conditions influencing fouling

The conditions influencing fouling can be classified as: (A) operating parameters, (B) heat exchanger parameters, and (C) fluid properties. Among the operating parameters the important events which influencing fouling at a significant level are: (1) velocity, (2) surface temperature, and (3) bulk temperature.

Velocity influences fouling at a significant level. In diffusion controlled processes, increasing the fluid velocity causes more fouling [24]. In most cases, fouling decreases at higher fluid velocities [4, 13, 25]. Increasing flow velocity increases the fluid shear stress which causes more removal. This results in lower fouling rates which resulting to lower fouling resistance. For weak deposits (particulate fouling), increasing the flow velocity may completely eliminate fouling. For stronger deposits, increasing the flow velocity beyond a particular point may not decrease fouling significantly [25]. For very strong deposits, increasing the flow velocity may not have any effect at all [6].

Surface temperature may increase, decrease or have no effect on fouling [26]. The rates of chemical reaction and inverse solubility crystallization increase with an increase in temperature. For inverse solubility salts, higher surface temperature increases fouling due to higher concentration gradients and higher reaction rate constants. In case of normal solubility salts cooling results in more fouling.

The bulk temperature also effects on increase of fouling rate. In inverse crystallisation, when precipitation happens in the fluid bulk, increasing the temperature increases the rate of crystal formation and hence deposition. Thus bulk temperature has effects on chemical reaction rate and polymerisation rate.

The important heat exchanger parameters are classified as: surface material, surface structure (roughness), heat exchanger type and geometry [27]. Surface material is considered seriously for corrosion fouling because of the potential to react and form corrosion products. Different materials have different catalytic action and may promote or reduce fouling for different processes. The initial fouling rate and scale formation depends significantly on the surface roughness. Junghahn [28] proved theoretically that the free energy change associated with crystal nuclei formation was much less on a rough surface than on a smooth surface. Rough surfaces result in higher deposition due to protected zones in the cavities or pits where flow velocities are very low.

According to Rankin and Adamson [29], it is not the rate of nucleation but the nuclei attachment which is strongly dependent on the surface roughness. Chandler [30] also observed similar results. In general the rough surface causes more fouling which reduces the delay time for all types of fouling. Surface roughness increases turbulence near the surface, which in turn increases the removal rate of fouling on the surface. Better performance occurred due to the increase in surface roughness with deposit formation and has been reported by some authors [1, 2]. Marriott [31] reiterated that mirror finished surfaces in heat exchangers are used to reduce fouling in practice.

5. Heat exchanger type, geometry and process fluid influencing fouling

Shell and tube heat exchangers are used most commonly but they are not particularly suitable for fouling conditions. Fouling can be reduced with special baffle and tube design. Several studies [32-35] have shown that finned tubes foul less than plain tubes. Non-uniform thermal expansion leads to lower deposit strength and hence less deposition. Freeman et al. [36] found that tubes with longitudinal grooves on the outside had less particulate fouling (by alumina particles) than the plain tubes.

Fluidised-bed heat exchangers are used in several applications to reduce or even eliminate fouling completely. Fluidised particles remove deposits from the heat transfer surface. They also enhance the heat transfer efficiency as they interrupt the viscous sub-layer. These heat exchangers have been used successfully to reduce fouling by hard, adhering silica deposits [37]. Graphite heat exchangers are also reported to have less fouling. Direct contact heat transfer may be another alternative to reduce fouling [38]. Properties of the process fluid such as the nature and concentration of the dissolved constituents or suspended particles, presence of any living organisms, solution pH etc. affect fouling significantly.

Excessively high over-concentration of solids in the evaporating liquid may lead to carry-over in the steam and cause fouling in process heat transfer equipment. Corrosion is very important on the steam side of process equipment. Water pH, over-concentration of treatment chemicals in evaporating liquids and dissolved gases (mainly oxygen and carbon dioxide) are very important contributors to corrosion fouling [39]. The presence of living organisms causes biological fouling and makes biofilms. This can sometimes enhance other fouling mechanisms too, as microbial deposits may trap suspended particles. They may also change the chemistry of water and can cause scaling or corrosion [39].

6. Fouling models

A number of models have been proposed for different types of fouling. Analysis and model improvement is still progressing as there are difficulties due to the complex nature of deposit formation and lack of reproducible measurement of fouling resistance. Most of the models have been simplified with many assumptions [40] as stated below:

- Surface roughness is neglected.
- Change in surface roughness with deposit formation is also neglected.
- Only one type of fouling is usually considered.
- Changes in physical properties of the fluids are neglected in most of the cases.
- The fouling layer is assumed to be homogeneous.

- Changes in flow velocity with changing cross-sectional area due to fouling are usually neglected.
- The shape of deposits, e. g. crystals or particles is ignored.

It is also observed that few attempts have been made to model the initiation or roughness delay period. Almost all the models predict fouling (scaling) after the delay period. Some other notable parameters are neglected in modelling such as: (a) effect of simultaneous action of different fouling mechanisms, (b) equipment design, (c) surface parameters e.g. surface material and surface roughness, (d) increase in surface area with deposition, (e) properties of foulant stream, (f) nature of process, and the (g) fluctuations in operation.

Modelling is usually done taking into consideration only (a) flow velocity, (b) concentration, (c) wall and bulk temperature, and (d) time.

Watkinson and Martinez [11] developed a model, based on the fundamental material balance equation (2.1). For the deposition rate the following expression is adopted:

The deposition rate is expressed as shown in equation (6.1).

$$\frac{dx_f}{dt} = \frac{K_R}{\rho_f}(c_F - c_{Sa})^n \tag{6.1}$$

For sparingly soluble salts with inverse solubility (e.g. $CaCO_3$), the deposition rate is controlled by the slow reaction rate and the constant of reaction rate K_R that obeys the Arrhenius equation:

$$K_R = A_0 e^{(\frac{-E}{R_g T_f})} \tag{6.2}$$

with T_f as the interface temperature.

Kern and Seaton [43] recommend for the removal rate the equation:

$$\dot{m}_r = a_8 \tau_f x_f \tag{6.3}$$

Where τ_f is the shear stress exerted by the liquid flow on the fouling film. Even though $CaCO_3$ deposits are much stronger than the particulate deposits considered by Kern and Seaton [43] the removal rate was assumed to be directly proportional to deposit thickness, which may not be correct for all the cases.

Kern and Seaton [41] proposed a model for particulate fouling which takes into account removal or re-entrainment of deposits. The mathematical model is based on a general material balance equation (2.1). Deposition and removal rates act separately and combine into a net deposition rate. The rate of deposition is expressed as:

$$\dot{m}_d = a_9 c' w \tag{6.4}$$

Where, c' is dirt concentration and w is constant weight flow of fluid. The removal rate is roughly proportional to the total depth of dirt deposited on the heat transfer surface as stated below.

$$\dot{m}_r = a_{10}\tau_f x_f \tag{6.5}$$

Combining the equations for deposition and removal rates (6.4) and (6.5) with the material balance equation (2.1), the fouling resistance expression is obtained:

$$R_f = \overset{*}{R}_f(1 - e^{-\theta t}) \tag{6.6}$$

where θ is a time constant and $\overset{*}{R}_f$ is the asymptotic value of the fouling resistance. For these also the following equations are obtained.

$$\overset{*}{R}_f = \frac{a_9 c'w}{a_{10}\lambda_f\tau_f} \tag{6.7}$$

$$\theta = a_{10}\tau_f \tag{6.8}$$

Here, λ_f is the thermal conductivity of the deposits, a_9 and a_{10} are proportionality constants. This model predicts asymptotic fouling behaviour with $\overset{*}{R}_f$ being the fouling resistance after an infinite time of operation. According to this model, no matter what the conditions, i.e. type of fluid, heat exchanger surface, temperature driving force, an asymptotic fouling value will be obtained sooner or later with removal rates becoming equal to deposition rates.

7. Cost imposed due to fouling

An additional cost is imposed by fouling of heat transfer equipment in industries. Few studies have been undertaken to determine the fouling related costs in industry. Fouling costs can generally be divided into four major categories, such as (1) increased capital expenditure, (2) energy costs, (3) maintenance costs, (4) cost of production loss and (v) extra environmental management cost.

Country	Fouling costs US $ million	GNP (1984) US $ million	Fouling costs % of GNP
USA (1982)	3860-7000	3634000	0.12-0.22
	8000-10000		0.28-0.35
Japan	3062	1225000	0.25
West Germany	1533	613000	0.25
UK (1978)	700-930	285000	0.20-0.33
Australia	260	173000	0.15
New Zealand	35	23000	0.15
Total Industrial World	26850	13429000	0.20

Table 7.1. Estimated fouling costs incurred in some countries.

The heat transfer area of a heat exchanger is kept exaggerated to compensate retardation imposed by fouling. Oversized pumps and fans are selected to compensate design over-surfacing the enhanced pressure loss from reduction in the flow area.

In some occasions standby heat exchangers are kept in process design in order to ensure uninterrupted operation while a fouled heat exchanger is taken under cleaning maintenance. In-situ cleaning in some cases are recommended while chemical cleaning is preferred for others. All together, cost of cleaning, cleaning equipment, chemicals all are imposing extra to the capital cost of the plant.

Muller-Steinhagen [37] reported that total annual costs for highly industrialised countries such as the United States and the United Kingdom are about 0.25 percent of the countries gross national product (GNP). Even for a relatively less industrialised country like New Zealand, the total fouling costs are around 0.15 percent of its GNP. Muller-Steinhagen [37] has summarised the total fouling costs for various countries based on 1984 in Table 7.1.

8. Fouling mitigation

Gilmour [42] reported that the degradation of heat transfer performance due to fouling in shell and tube heat exchangers occurs mainly due to poor shell-side design. In recent years numerous methods have been developed to control fouling. These methods can be classified as: (1) chemical methods, (2) mechanical methods and (3) changing the phase of the solution. By adding foreign chemicals in a solution, reduction of fouling is achieved by chemical methods of fouling mitigation. Chemical additives developed by many companies have been extensively used to mitigate fouling in the industrial sector. Various additives can be used to prevent scaling [43-44]. Bott [45] specified that the additives used act in different ways, such as (a) sequestering agents, (b) threshold agents, (c) crystal modifiers and (d) dispersants. Some of the common water additives are EDTA (sequestering agent), polyphosphates and polyphosphonates (threshold agents) and polycarboxylic acid and its derivatives (sequestering and threshold treatment). Sequestering agents such as EDTA complex strongly with the scaling cations such as Ca^{++}, Mg^{++}, and Cu^{++} in exchange with Na^+, thus preventing scaling as well as removing any scale formed previously. They are used effectively as antiscalants in boiler feed water treatment. Troup and Richardson [46] claimed that their use is uneconomical when hardness levels are high.

Polyphosphates and polyphosphonates as threshold agents are also used to reduce scaling in boilers and cooling water systems. Bott [45] said that they prevent the formation of nuclei thus preventing the crystallisation and mitigate fouling. Very small quantities of these agents are effective in reducing scaling from supersaturated salt solutions.

Crystal modifying agents (e.g. Polycarboxylic acid) distort the crystal habit and inhibit the formation of large crystals. The distorted crystals do not settle on the heat transfer surface, they remain suspended in the bulk solution. If their concentration increases beyond a certain limit, particulate fouling may take place. This is prevented either by using techniques to minimise particulate fouling or using dispersing agents along with crystal modifying agents.

Though crystallisation fouling may not be prevented completely using additives, the resulting crystalline deposits are different from those formed in the absence of any additives. The layer looses its strength and can be removed easily. By controlling pH, crystallisation fouling can furthermore be minimised. The solubility of deposit forming components usually increases with decreasing pH. In many water treatment plants, sulphuric acid is added to maintain a pH between 6.5 and 7.5 [47]. In this case, addition of corrosion inhibitors may also be required which may enhance fouling again.

Seeding is used commercially to reduce crystallisation fouling. This method involves addition of seeds to the scaling fluid. Crystallisation takes place preferentially on these seeds rather than on the heat transfer surface. Calcium sulphate seeds are generally used to avoid calcium sulphate scaling [48-49]. These seeds need not be of the crystallising material, but they should have similar crystallographic properties, i. e. atomic agreement and lattice spacing [50].

To mitigate particulate fouling by chemical means, dispersants are used to reduce the surface tension of deposits. It helps in disintegrating the suspended particles into smaller fragments that do not settle so readily.

Addition of certain chemicals can slow down or terminate chemical reactions. Dispersants are very helpful in keeping the foulants away from the surface. Some particles such as corrosion products may act as catalysts. Chemical reaction fouling could be suppressed by reducing the number of these particles. Corrosion inhibitors (chromates and polyphosphates) can be used to reduce corrosion fouling [47]. Usually a passivating oxide layer is desired to prevent corrosion of the surface. Corrosion fouling may promote other fouling mechanisms e. g. higher roughness of the corroded surface may enhance crystallisation fouling. The corrosion products may act as catalysts and promote chemical reaction fouling and also augments particulate fouling by depositing on the heat transfer surface.

Mitigation of fouling by chemical methods has several drawbacks. Fouling and corrosion inhibitors usually contain considerable amount of chlorine, bromine, chromium, zinc etc. Therefore, their concentration has to be monitored carefully. Treatment of fluid released from the plant to natural waterways is necessary to prevent harmful effects. Higher concentrations can be used in closed systems but overdosing may have negative effects and some components may precipitate. Using different additives at the same time may result in dangerous chemical reactions. Some additives have limited life and some degrade with time and loose activity.

Pritchard [51] has broadly classified mechanical methods into two categories according to their ways of action. (1) Brute force methods such as high-pressure jets, lances, drills etc. (2) Mild methods such as brushes and sponge balls. Muller-Steinhagen [37] has reported that several mechanical methods have been developed in recent years. The following mechanisms predict the modern methods:

- Breakage of deposits during brief overheating due to differential thermal expansions of heat transfer surface and deposits,
- Mechanical vibration of the heat transfer surfaces,
- Acoustical vibration of the surface,
- Increased shear stress at the fluid/deposit interface, and
- Reduced stickiness of the heat transfer surface.

Most liquid-side fouling mitigation techniques have been developed for the tube-side of shell and tube heat exchangers. The relevant techniques include:

1. increase in flow velocity,
2. reversal of flow direction,
3. heat transfer surface such as, surface roughness and surface materials,
4. fluidised bed heat exchangers,
5. pulsating flow,
6. turbulence promoters, and
7. transport of cleaning devices through tubes.

The deposits which are not strongly adhere to the surface can be removed by increasing the flow velocity. Muller-Steinhagen and Midis [4] reported that alumina deposits were removed completely when the flow velocity was increased for a short period of time after a fouling run. At higher flow velocity, the wall shear stress increases and causes more removal of deposits from the surface.

At a regular interval of time, the reversal of flow direction on the heat transfer surface could be another effective method of reducing fouling. This technique needs several modifications in the existing set-up. Muller-Steinhagen [37] stated that mitigation of fouling by increasing the flow velocity could be more effective than reversal of flow direction.

Surface material and surface roughness play an important role on fouling mitigation. Thus lowering the surface roughness retards the adhesion of deposits and the number of nuclei growth sites. Lower deposition rate also experienced with lowering surface energy of the material of heat exchanger. Using inert particles is an effective way of reducing or even eliminating fouling completely as practiced in fluidized bed heat exchangers. Pulsating flow in heat exchangers is a strategy to increase the level of turbulence [52-58]. Where, as a matter of fact heat transfer coefficient increases with the enhancement of deposit removal. Higher heat transfer reduces fouling by reducing the interface temperature which is beneficial for certain fouling mechanism such as crystallization fouling of inverse solubility salts. The higher level of turbulence augments the deposit removal rate.

Fracture of deposits by fatigue is enhanced by higher turbulence due to pulsation resulting to increase of removal rate. Generally the deposition rate of fouling phenomena [3] depends on the thickness of viscous and thermal sub-layers. Muller-Steinhagen [37] reported that by inserting turbulence promoters inside tubes or by using tube corrugations, the heat transfer coefficient can be increased by a factor of 2 to 15 by reducing the thickness of average thermal boundary layer. Turbulence promoters may reduce both the crystallisation and reaction fouling. Muller-Steinhagen [37] informed that particulate fouling will be enhanced if particulate or fibrous material already exists in the solution.

Middis [10] also reported fouling mitigation by adding natural fibre in the supersaturated solutions of concentration 3.6 g/L $CaSO_4$. He observed that the rate of $CaSO_4$ fouling on heated metal tube surface decreases with the increase of fibre concentration in the fouling solution. Kazi [59] also got similar results by adding different types and concentrations of natural fibre in supersaturated solutions of $CaSO_4$.

Some novel methods which do not fall under well reported categories, such as magnetic or electric treatment are also available in the market to reduce fouling. Usually magnetic treatment is carried out by inserting permanent magnets in a pipe before the heat exchanger.

Parkinson and Price [60] have reported significant reduction in fouling by the magnetic treatment as it helps in precipitating the salts. These salts stay suspended in the bulk liquid and are removed later. On the other hand Hasson and Bramson [61] informed that there is no effect of magnetic treatment at all on fouling. They observed that magnetic treatment neither decreased nor increased the rate of scaling. The nature of the deposits also remained unchanged. Bernadin and Chan [62] have also reported no influence of magnetic treatment on silica fouling. Muller-Steinhagen [37] has stated that magnetic mitigation devices in some cases actually increased fouling. Thus from the available information no conclusion can be made about the influence of the magnetic field on the scaling process.

9. Cleaning of heat exchangers

A decrease in the performance of a heat exchanger beyond acceptable level requires cleaning. In some applications, the cleaning can be done on line to maintain acceptable performance without interruption of operation. At other times, off-line cleaning must be used.

Garrett-Price et al. [27] presented some cleaning approaches for fouled heat exchangers. They specified on-line cleaning generally utilises a mechanical method designed for only tube side and requires no disassembly. In some applications flow reversal is required. Chemical feed can also be used as an on-line cleaning technique but may upset the rest of the liquid service loop.

On-line mechanical cleaning techniques are also in practice. On line tube side cleaning techniques are the sponge-ball and brush systems. The advantage of on-line cleaning is the continuity of service of the exchanger and the hope that no cleaning-mandated downtime will occur. The principal disadvantage is the added cost of a new heat exchanger installation or the large cost of retrofits. Furthermore there is no assurance that all tubes are being cleaned sufficiently.

Off-line chemical cleaning is a technique that is used very frequently to clean exchangers. Some refineries and chemical plants have their own cleaning facilities for dipping bundles or re-circulating cleaning solutions. In general, this type of cleaning is designed to dissolve the deposit by means of a chemical reaction with the cleaning fluid. The advantages of chemical cleaning approach include the cleaning of difficult-to-reach areas. Often in mechanical cleaning, there is incomplete cleaning due to regions that are difficult to reach with the cleaning tools. There is no mechanical damage to the bundle from chemical cleaning, although there is a possibility of corrosion damage due to a reaction of the tube material with the cleaning fluid. This potential problem may be overcome through proper flushing of the unit. Disadvantages of off-line chemical cleaning include corrosion damage potential, handling of hazardous chemicals, use of a complex procedure.

Off-line mechanical cleaning is a frequently used procedure. The approach is to abrade or scrap away the deposit by some mechanical means. The method includes high-pressure water, steam, lances and water guns. In off-line mechanical cleaning there are some advantages such as excellent cleaning of each tube is possible, good removal potential of very tenacious deposits. Disadvantages include the inability to clean U-tube bundles successfully, usual disassembly problem and the great labour needed.

Frenier and Steven [63] describe general methods for cleaning heat exchanger equipment, including both mechanical and chemical procedures. They have given guidelines for selecting between chemical and mechanical cleaning, and among the various types of chemical cleaning processes. They stated that water-based fluids can transport and deposit a wide variety of minerals, and corrosion products form due to the reaction of the aqueous fluids with the metals of construction. Hydrocarbon and petrochemical fluids transport and deposit a variety of organic scales. Common inorganic scale forming compound includes various iron oxides, hardness deposits (carbonates and silicates). They stated that the entire cleaning situation must be considered when choosing between mechanical and chemical cleaning, as well as the specific technique within the general category. The general categories of mechanical cleaning are abrasive, abrasive hydraulic, hydraulic and thermal [64].

Frenier and Barber [63] stated that, for chemical cleaning of the heat exchanger tubes, it is very beneficial to obtain a sample of the deposit so that its composition can be determined. Based on the chemical analysis of the deposit, an optimal treatment plan can be developed and the best solvent selected. They have classified the deposits generically, as organic (process-side) or inorganic (water-side).

They stated that the process side deposits may range from light hydrocarbon to polymers and generally they are similar to the fluids from which they originate. They mentioned that the general categories of solvents for process side scales include aqueous detergent solutions, true organic solvents and emulsions. Aqueous detergent formulations always contain a surfactant-type component. In addition they can contain alkaline agents, such as sodium hydroxide, sodium silicate, or sodium phosphate. Builder molecules such as ethylenediaminetetraacetic acid (EDTA) suppress the effects of hard water, and coupling agents such as glycol ethers, improve the dissolution of some organic deposits.

Detergent formulations are effective only for removing the lighter deposits. Refinery fluids, aeromatics and terpenes are used to dissolve the organic deposits. N-methyl-2-pyrrolidinone also is a very effective polar solvent with low toxicity characteristics. They reiterated that the effectiveness of the application depends greatly on proper application conditions, such as flow rate and temperature. Combination of surfactants, organic solvents and water emulsions are good cleaning agents. Emulsions with an organic outer phase are particularly useful for cleaning large vessels. Oily rust deposits having both organic and inorganic compositions can be removed by acidic emulsions combining an acid and an organic solvent.

Water-side deposits usually contain minerals, such as iron oxides (corrosion products), hardness (Ca and Mg carbonates) and silica, in individual cases other minerals can also be found. The solvents for removing inorganic deposits usually contain mineral acids, organic acids or chelating agents.

Mineral acids used in chemical cleaning include hydrochloric acid (HCl), hydrofluoric acid (HF), sulphuric acid (H_2SO_4), phosphoric acid (H_3PO_4), nitric acid (HNO_3) and sulfamic acid (H_2NSO_3H). Hydrochloric acid is the most common and most versatile mineral acid. It is used on virtually all types of industrial process equipment at strengths from 5 percent to 28 percent (5-10 percent is the most usual range). It can be inhibited at temperatures up to about 180 °F. HCl will dissolve carbonates, phosphates, most sulphates, ferrous sulphide, iron oxides and copper oxides. By using with appropriate additives, fluoride deposits, copper and silica can also be removed from surfaces with inhibited HCl. HCl is corrosive, so it has restricted use. HCl is not used to clean series 300 SS, free-machining alloys, magnesium, zinc, aluminium, cadmium, or galvanised steel because of the potential for generalised or localised attack. It is not desirable to contact the fouled metal with a strong mineral acid, because of the danger of damage to the equipment during or after cleaning. An alternative solvent family consists of aqueous solutions of chelating agents and organic acids with pH values of about 2 to 12.

Citric acid was one of the first organic acids used in industrial cleaning operations [65]. For removing iron oxide from steel surfaces, citric acid and a mixture of formic and citric acid could be used [66]. The mixture could hold more iron in solution than either of the acids alone could do. Ammonium citrate and sodium citrate solvents are currently used to clean a wide variety of heat transfer equipment, including boilers and various types of service water systems. The advantage of citric acid formulation is their low toxicity and ready biodegradability. EDTA is a versatile chemical that forms metal ion complexes with higher equilibrium constants than citric acid. As a result chemical cleaning solvents with pH values from 4.5 to about 9.2 have been formulated that can remove Fe and Cu, as well as Ca, Ni and Cr. The major advantage of the EDTA solvents is that they are much more aggressive than citric salts for removing very heavy iron oxide deposits especially if they contain copper. The disadvantage includes high cost per pound of metal removed and low biodegradability.

All of the chelating agents are also organic acids. Eberhard and Rosene [67] taught the use of solvents consisting of formic acid or citric acid for cleaning nondrainable tubes in super heaters. Reich [66] used a mixture of formic acid and citric acid to a proportion of 3:1, to remove iron oxide deposits. The advantage of these mixtures is that they avoid the precipitation of solids that formed in pure formic or citric acid solutions. Formulations of formic acid with hydroxyacetic acid and citric acid with hydroxyacetic acid can be used as a cleaning agent. Bipan [3] used acetic acid of concentration 3 percent to remove $CaSO_4.2H_2O$ deposit on plate type SS heat exchangers. He said that with the increase in acid solution temperature the removal efficiency increases. Similar results were obtained by Kazi [68]. It reveals that a complete and systematic study of fouling on different metal surfaces and their mitigation by additives have been required to be done along with study of introducing a benign to environment technique for chemical cleaning of fouling deposits.

10. Nomenclature

A	Heat transfer area	m^2
A_0	Arrhenius constant	$m^3/kg.s$
a_1-a_{13}	Proportionality constant	-
c	Concentration	g/L or kg/m^3
c_p	Specific heat capacity	$J/molK$
d	Pipe diameter	m
E	Activation energy	J/mol
ΔH	Head loss	$m\ H_2O$
h_c	Heat transfer coefficient	W/m^2K
K_R	Reaction rate constant (dimension depend on the order of n)	$m^4/kg.s$
L	Length	m

$\overset{\bullet}{m}$	Mass flux	kg/m²s
$\overset{\bullet}{m_d}$	Increase of solids mass present in the fouling film	kg/m²s
$\overset{\bullet}{m_r}$	Decrease of solids in the fouling film	kg/m²s
m_f	Solids deposited in the fouling film per unit area	kg/m²
P	Pressure	kPa
P	Perimeter	m
P_c	Intercrystalline cohesive force	N
ΔP	Pressure drop	kPa/m
$\overset{\bullet}{Q}$	Heat flow	W
$\overset{\bullet}{q}$	Heat flux	W/m²
R	Ratio of the radius of inner and outer tubes of annulus	-
R_b	Bonding resistance	-
R_g	Universal gas constant	J/mol.K
R_f	Fouling resistance	m²K/kW
$\overset{*}{R_f}$	Asymptotic value of the fouling resistance	m²K/kW
r	Radius	m
r_H	Hydraulic radius	m
T	Temperature	°C
T_f	Temperature at the surface of the fouling film	°C
ΔT	Temperature difference	K or °C
t	Time	s
t_{ind}	Induction time	s
U	Overall heat transfer coefficient	W/m²K
u	Velocity	m/s
$\bar{u}$	Local mean velocity	m/s
u*	Friction velocity, $\sqrt{(\tau_w/\rho)}$	m/s
u⁺	Dimensionless velocity, $\bar{u}/u*$	-
u_t	Turbulent friction or shear velocity, $\bar{u}\sqrt{f/2}$	m/s
w	Constant weight flow of fluid	kg/s
x	Distance in x direction	m
x_f	Fouling film thickness	m
y	Distance in y direction	m

Greek

α	Constant	-
β	Individual mass transfer coefficient	m/s
θ	Time constant	s
ε	Height of roughness	m

ε/d	Roughness ratio	-
f	Fanning friction factor	-
λ	Thermal conductivity	W/mK
λ_f	Thermal conductivity of the deposits	W/mK
μ	Dynamic viscosity	kg/ms
ρ	Density	kg/m³
ρ_f	Density of the deposits	kg/m³
τ	Shear stress	N/m²
τ_w	Wall shear stress	N/m²
τ_f	Shear stress exerted by the liquid flow on the fouling film	N/m²
ν	Kinematic viscosity	m²/s
ϕ	Friction factor	-
δ	Hydrodynamic boundary layer thickness	m
δc	Linear thermal expansion coefficient of the fouling film porosity	1/K
δ_t	Thermal boundary layer thickness	m
ξ	Ratio of thermal to hydrodynamic boundary layers	m

Dimensionless Numbers

Nusselt Number $\qquad Nu = \dfrac{h_c \cdot d}{\lambda}$

Prandtl Number $\qquad Pr = \dfrac{c_p \cdot \mu}{\lambda}$

Reynolds Number $\qquad Re = \dfrac{\rho \cdot u \cdot d}{\mu}$

11. References

[1] Bott, T. R. and Gudmundsson, J. S., Rippled Silica deposits in Heat Exchanger Tubes. 6th International Heat Transfer Conference. 1978.

[2] Crittenden, B. D. and Khater, E. M. H., Fouling From Vaporising Kerosine. Journal of Heat Transfer, 1987. 109: p. 583-589.

[3] Bansal, B., Crystallisation Fouling in Plate Heat Exchangers, PhD thesis, Department of Chemical and Materials Engineering, 1994, The University of Auckland: Auckland, New Zealand.

[4] Muller-Steinhagen, H. M. and Middis, J., Particulate Fouling in Plate Heat Exchangers. Heat Transfer Engineering, 1989. 10(4): p. 30-36.

[5] Bohnet, M., Fouling of Heat Transfer Surfaces. Chemical Engineering Technology, 1987. 10: p. 113-125.

[6] Ritter, R. B., Crystallisation Fouling Studies. Journal of Heat Transfer, 1983. 105: p. 374-378.

[7] Reitzer, B. J., Rate of Scale Formation in Tubular Heat Exchangers. I & EC Process Design and Development, 1964. 3(4): p. 345-348.

[8] Hasson, D., Rate of Decrease of Heat Transfer Due to Scale Deposition. DECHEMA Monograph, 1962. 47: p. 233-282.

[9] Bott, T. R. and Walker, R. A., Fouling in Heat Transfer Equipment. The Chemical Engineer, 1971: p. 391-395.

[10] Middis, J., Heat Transfer and Pressure Drop For Flowing Wood Pulp Fibre Suspensions, PhD thesis, Chemical and Materials Engineering. 1994, The University of Auckland: Auckland, New Zealand.

[11] Watkinson, A. P. and Martinez, O., Scaling of Heat Exchanger Tubes by Calcium Carbonate. Journal of Heat Transfer, 1975: p. 504-508.

[12] Augustin, W., Verkrustung (Fouling) Von Warmeubertragungsflachen, in Institut fur Verfahrens- und Kerntechnik. 1992, Technische Universitat Braunschweig: Germany.

[13] Cooper, A., Suitor, J. W. and Usher, J. D., Cooling Water Fouling in Plate Heat Exchangers. Heat Transfer Engineering, 1980. 1(3): p. 50-55.

[14] Muller-Steinhagen, H. M., Reif, F., Epstein, N. and Watkinson, A. P., Influence of Operating Conditions on Particulate Fouling. The Canadian Journal of Chemical Engineering, 1988. 66: p. 42-50.

[15] Bott, T.R., Fouling of Heat Exchangers. 1995: Elsevier Science & Technology Books. 529.

[16] Bell, K.J. and A.C. Mueller, Wolverine Heat Transfer Data book II. 2001, Wolverine Tube, Inc.

[17] Epstein, N., Heat Exchanger Theory and Practice, in: J. Taborek, G. Hewitt (eds.) heat exchangers in Heat Exchanger Theory and Practice, McGraw-Hill, 1983.

[18] Epstein, N., Thinking about Heat transfer fouling: a 5 x 5 matrix. heat Transfer Engineering, 1983. 4: p. 43-46.

[19] Epstein, N. (1981) Fouling in heat exchangers. In Low Reynolds Number Flow Heat Exchangers, S. Kakac, R. K. Shah, and A. E. Bergles (eds.). Hemisphere, New York.

[20] Somerscales, E. F. C., and Knudsen, J. G. (eds.) (1981) Fouling of Heat Transfer Equipment. Hemisphere, New York.

[21] Melo, L. F., Bott, T. R., and Bernardo, C. A. (eds.) (1988) Fouling Science and Technology. Kluwer, Dordrecht.

[22] Standards of the Tubular Exchanger Manufacturers Association 7th ed. Tubular Exchanger Manufacturers Association, New York, 1988.

[23] Chenoweth, J. M. (1988), General design of heat exchangers for fouling conditions. In Fouling Science and Technology, L. F. Melo, T. R. Bott, and C. A. Bernardo (eds.), pp. 477-494. Kluwer, Dordrecht.

[24] Brusilovsky, M., Borden, J. and Hasson, D., Flux Decline due to Gypsum Precipitation on RO Membranes. Desalination, 1992. 86: p. 187-222.

[25] Walker, G., Degradation of Performance, Industrial Heat Exchangers- A Basic Guide. 1982: Hemisphere Publishing Corporation. 213-272.

[26] Gudmundson, J. S., Particulate Fouling, in Fouling of Heat Transfer Equipment, E.F.C. Somerscales and J.G. Knudsen, Editors. 1981, Hemisphere Publishing Corporation. p. 357-387.

[27] Garrett- Price, B. A., Smith, S. A., Watts, R. L., Knudsen, J. G., Marner, W. J. and Suitor, J. W., Overview of Fouling, Fouling of Heat Exchangers- Characteristics, Costs, Prevention, Control, and Removal. 1985, Noyes Publications: New Jersey. p. 9-20.

[28] Junghahn, L., Methoden Zum Herabsetzen oder Verhindern der Krustenbildung. Chemie Ingenieur Technik, 1964. 36: p. 60-67.

[29] Rankin, B. H. and Adamson, W. L., Scale Formation as Related to Evaporator Surface Conditions. Desalination, 1973. 13: p. 63-87.

[30] Chandler, J. L., The Effect of Supersaturation and Flow Conditions on the Initiation of Scale Formation. Transactions of Institution of Chemical Engineers, 1964. 42: p. T24-T34.

[31] Marriott, J., Where and How to Use Plate Heat Exchangers. Chemical Engineering, 1971. 78(8): p. 127-134.

[32] Knudsen, J. G. and McCluer, H. K., Hard Water Scaling of Finned Tubes at Moderate Temperatures. Chem. Eng. Progress Symp. series, 1959. 55(9): p. 1-4.

[33] Katz, D. L., Knudsen, J. G., Balekjian, G. and Grover, S. S., Fouling of Heat Exchangers. Petroleum Refiner, 1954. 33(4): p. 121-125.

[34] Webber, W. O., Under Fouling Conditions Finned Tubes can Save Money. Chemical Engineering, 1960. 67(6): p. 149-152.

[35] Sheikholeslami, R. and Watkinson, A. P., Scaling of Plain and Externally Finned Tubes. Journal of Heat Transfer, 1986. 108: p. 147-152.

[36] Freeman, W. B., Middis, J. and Muller-Steinhagen, H. M., Influence of Augmented Surfaces and of Surface Finish on Particulate Fouling in Double Pipe Heat exchangers. Chemical Engineering Processing, 1990. 27: p. 1-11.

[37] Muller-Steinhagen, H. M., Fouling: The Ultimate Challenge for Heat Exchanger Design. The sixth International Symposium on Transport Phenomena in Thermal Engineering. 1993. Seoul, Korea.

[38] Muller-Steinhagen, H. M., Introduction to Heat Exchanger Fouling. Proceedings of Fouling in Heat Exchangers. 1988. The University of Auckland, Centre for Continuing Education, Auckland, New Zealand.

[39] Garrett- Price, B. A., Smith, S. A., Watts, R. L., Knudsen, J. G., Marner, W. J. and Suitor, J. W., Generic Industrial Fouling, in Fouling of Heat Exchangers- Characteristics, Costs, Prevention, Control, and Removal. 1985, Noyes Publications. p. 21-37.

[40] Pinheiro, J. D. D. R. S., Fouling of Heat Transfer surfaces, Heat Exchanger Source Book, J.W. Palen, Editor. 1986, Hemisphere Publishing Corporation. p. 721-744.

[41] Kern, D. Q. and Seaton, R. E., A Theoretical Analysis of Thermal surface Fouling. Brit. Chem. Eng., 1959. 4(5): p. 258-262.

[42] Gilmour, C. H., No Fooling- No Fouling. Chem. Eng. Progr., 1965. 61(7): p. 49-54.

[43] Harris, A. and Marshall, A., The Evaluation of Scale Control Additives, Conference on Progress in the Prevention of Fouling in Industrial Plant. 1981. University of Nottingham.

[44] Krisher, A. S., Raw Water Treatment in the CPI. Chemical Engineering, 1978: p. 79-98.

[45] Bott, T. R., The Fouling of Heat Exchangers. DSIR, Wellington, New Zealand., 1981.

[46] Troup, D. H. and Richardson, J. A., Scale Nucleation on a Heat Transfer Surface and its prevention. Chemical Engineering Communications, 1978. 2: p. 167-180.

[47] Muller-Steinhagen, H. M., Fouling of Heat Transfer Surfaces. VDI Heat Atlas, English Edition, VDI- Verlag GmbH, 1993: p. OC1-OC22.

[48] Gainey, R. J., Thorp, C. A. and Cadwallader, Calcium Sulphate Seeding Prevents Calcium sulphate Scaling. Industrial and Engineering Chemistry, 1963. 55(3): p. 39-43.

[49] Rautenbach, R. and Habbe, R., Seeding Technique for Zero-Discharge Processes, Adaption to Electrodialysis. Desalination, 1991. 84: p. 153-161.

[50] Telkes, M., Nucleation of Supersaturated Inorganic Salt Solutions. Industrial and Engineering Chemistry, 1952. 44(6): p. 1308-1310.

[51] Pritchard, A. M., Cleaning of Fouled Surfaces: A Discussion, in Fouling Science and Technology, NATO ASI Series, Series E: Applied Science, Melo, L. F., Bott, T. R. and Bernardo, C. A., Editors. 1988. p. 721-726.

[52] Keil, R. H., Enhancement of Heat Transfer by Flow Pulsation. Industrial Engineering Chemistry: Process Design and Development, 1971. 10(4): p. 473-478.

[53] Ludlow, J. C., Kirwan, D. J. and Gainer, J. L., Heat Transfer with Pulsating Flow. Chemical Engineering Communications, 1980. 7: p. 211-218.

[54] Karamercan, O. E. and Gainer, J. L., The Effect of Pulsations on Heat Transfer. Industrial Engineering Chemistry Fundamentals, 1979. 18(1): p. 11-15.

[55] Herndon, R. C., Hubble, P. E. and Gainer, J. L., Two Pulsators for Increasing Heat Transfer. Industrial Engineering Chemistry: Process Design and Development, 1980. 19: p. 405-410.

[56] Edwards, M. F. and Wilkinson, W. L., Review of Potential Applications on Pulsating Flow in Pipes. Transactions of Institution of Chemical Engineers, 1971. 49: p. 85-94.

[57] Gupta, S. K., Patel, R. D. and Ackerberg, R. C., Wall Heat /Mass Transfer in Pulsatile Flow. Chemical Engineering Science, 1982. 37(12): p. 1727-1739.

[58] Thomas, L. C., Adaptation of the surface Renewal Approach to Momentum and Heat Transfer for Turbulent Pulsatile Flow. Journal of Heat Transfer, 1974: p. 348-353.

[59] Kazi, S. N., Heat Transfer to Fibre Suspensions-Studies in Fibre Characterisation and Fouling Mitigation, PhD thesis, Chemical and Materials Engineering. 2001, The University of Auckland: Auckland, New Zealand.

[60] Parkinson, G. and Price, W., Getting the Most out of Cooling Water. Chemical Engineering, 1984: p. 22-25.

[61] Hasson, D. and Bramson, D., Effectiveness of Magnetic Water Treatment in Suppressing Calcium Carbonate Scale Deposition. Industrial Engineering Chemistry: Process Design and Development, 1985. 24: p. 588-592.

[62] Bernardin, J. D. and Chan, S. H., Magnetics Effects on Similated Brine Properties Pertaining to Magnetic Water Treatment. Fouling and Enhancement Interactions, 28th National Heat Transfer Conference. 1991. Minneapolis, Minnesota, USA.

[63] Frenier, W. W. and Barber, J. S., Choose the Best Heat Exchanger Cleaning Method. Chemical Engineering Progress, 1998: p. 37-44.

[64] Gutzeit, J., Cleaning of Process Equipment and Piping. 1997, MTI Publication, Materials Technology Institute, St. Louis.

[65] Loucks, C. M., Organic Acids for Cleaning Power- Plant Equipment. Annual Meeting of ASME. 1958. New York.

[66] Reich, C. F., Scale Removal. 1961: U. S. A.

[67] Eberhard, J. F. and Rosene, R. B., Removal of Scale Deposits. 1961: U. S. A.

[68] Kazi, S. N., Duffy, G. G. And Chen, X. D., Mineral scale formation and mitigation on metals and a polymeric heat exchanger surface. Applied Thermal Engineering, 30 (2010): p. 2236-2242.

7

Fouling in Plate Heat Exchangers: Some Practical Experience

Ali Bani Kananeh and Julian Peschel
GEA PHE Systems
Germany

1. Introduction

Due to their compact size, Plate Heat Exchangers (PHEs) are widely used in industrial processes. They have higher heat-transfer performance, lower temperature gradient, higher turbulence, and easier maintenance in comparison with shell and tube heat exchangers. For minimizing material consumption and space requirements compact models have been developed over the last years. By using thin plates forming a small gap, these compact models impress with larger heat transfer coefficients and, thus, smaller required heat transfer area.

The advantages of compact heat exchangers over shell and tube ones at a glance:

- larger heat transfer coefficients
- smaller heat transfer surfaces required
- lower fouling due to high fluid turbulences (self-cleaning effect)
- significantly smaller required installation and maintenance space
- lighter weight
- simplified cleanability especially for GPHE
- lower investment costs
- closer temperature approach
- pure counter-flow operation for GPHE

In Figure 1, plate heat exchangers are compared with shell and tube heat exchangers regarding effectiveness, space, weight and cleaning time.

Deposits create an insulating layer over the surface of the heat exchanger that decreases the heat transfer between fluids and increases the pressure drop. The pressure drop increases as a result of the narrowing of the flow area, which increases the gap velocity (Wang et al., 2009). Therefore, the thermal performance of the heat exchanger decreases with time, resulting in an undersized heat exchanger and causing the process efficiency to be reduced. Heat exchangers are often oversized by 70 to 80%, of which 30 to 50% is assigned to fouling. While the addition of excess surface to the heat exchanger may extend the operation time of the unit, it can cause fouling as a result of the over-performance caused by excess heat transfer area; because the process stream temperature change greater than desired, requiring that the flow rate of the utility stream be reduced (Müller-Steinhagen, 1999). The deposits

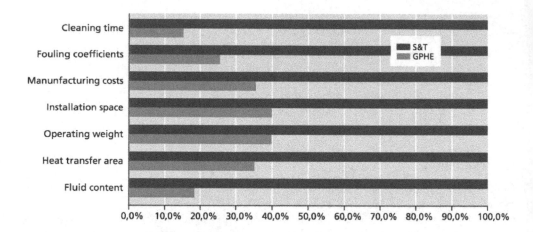

Fig. 1. Comparison of plate heat exchanger with shell and tube heat exchanger.

must be removed by regular and intensive cleaning procedures in order to maintain production efficiency.

As a result of the effects of fouling on the thermal and hydraulic performance of the heat exchanger, an additional cost is added to the industrial processes. Energy losses, lost productivity, manpower and cleaning expenses cause immense costs. The annual cost of dealing with fouling in the USA has been estimated at over $4 billion (Wang et al., 2009).

The manner in which fouling and fouling factors apply to plate exchangers is different from tubular heat exchangers. There is a high degree of turbulence in plate heat exchanger, which increases the rate of deposit removal and, in effect, makes the plate heat exchanger less prone to fouling. In addition, there is a more uniform velocity profile in a plate heat exchanger than in most shell and tube heat exchanger designs, eliminating zones of low velocity which are particularly prone to fouling. Figure 2 shows the fouling resistances for cooling water inside a plate heat exchanger in comparison with fouling resistances on the tube-side inside a shell and tube heat exchanger for the same velocity. A dramatic difference in the fouling resistances can be seen. The fouling resistances inside the PHE are much lower than that inside the shell and tube heat exchanger.

Fouling inside heat exchanger can be reduced by:

- Appropriate heat-exchanger design
- Proper selection of heat-exchanger type
- Mitigation methods (mechanical and/or chemical)
- Heat exchanger surface modification/coating

The mechanics of deposits build-up and the impact of operating conditions on the deposition rate should be understood in order to select the appropriate method to reduce fouling (Müller-Steinhagen, 1999).

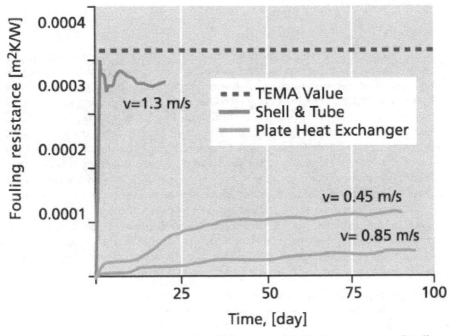

Fig. 2. Comparison of fouling resistance in PHE to tube-side fouling resistance (Müller-Steinhagen, 2006).

This chapter focuses on solving fouling problems in some industrial applications. The first section presents fouling problems with cooling water inside CO_2 coolers in different Egyptian fertilizer plants. The effect of heat exchanger geometry and flow patterns on the fouling behavior will be shown. Thermodynamic and hydraulic solutions are proposed like, redesigning of the plate heat exchangers and new plate geometries. The second section explains how fouling can be reduced inside gasketed plate heat exchangers used in food production using Nano-composite coatings. An antifouling coating with low surface energy (low wettability) can be used to avoid or minimize adhesion, improve process management, simplify cleaning processes with less resources and chemical use, and increase product reliability. The operational efficiency of the plant can be significantly improved and the intensity and frequency of cleaning can be substantially reduced.

2. Solving fouling problems by heat exchanger design modification

Fouling problems with cooling water inside CO_2 coolers in different Egyptian fertilizer plants were investigated. Thermodynamic and hydraulic solutions were proposed, which included redesign of the existing PHEs and new plate geometries. The main problems arose from the large surface margins required to meet pressure drop limits on the CO_2 side. Reducing the surface area of the heat exchanger increased the fluid velocity (shear stress from 5.31 to 10.84 Pa) inside the gaps and hence decreased fouling. Using computer-modeled plate geometries from the new technology (NT) series with larger gap velocities due to better fluid distribution over the plates could decrease fouling and increase the availability of fertilizer plants.

2.1 Introduction

To guarantee production reliability in the complex urea fertilizer manufacturing process, PHEs are installed in several process chains including CO_2 cooling, residual gas scrubbing, and other process sections as were as in the primary urea production plant. Industrial processes commonly use water for cooling purposes. Open circuit cooling system is used in some processes, while closed loop system involving cooling towers is used in others. Closed loop systems usually cause less fouling than open ones, but they are more expensive (Kukulka and Leising, 2009). Cooling water normally contains dissolved or suspended solids like calcium carbonate and calcium sulphate. If the concentration of these dissolved solids exceeds certain limits, it leads to the accumulation of deposits on the heat exchanger surface (Müller-Steinhagen, 1999). These deposits create an insulating layer on the surface of the heat exchanger that decreases the heat transfer between the two fluids. The thermal performance of the unit decreases with time as the thickness of the deposit increases, resulting in an undersized heat exchanger and causing the process efficiency to be reduced (Kukulka and Leising, 2009). Deposit formation can be reduced either by changing the configuration of the heat exchanger or by regular cleaning procedures.

Deposit formation is influenced by the heat exchanger surface and geometry, cooling medium and the operating conditions. Its composition depends on the flow rate, temperature and chemical composition of the cooling medium (Kukulka and Leising, 2009). Pana-Suppamassadu et al. (2009) studied the effect of plate geometry (contact angle) and the gap velocity on calcium carbonate fouling in plate heat exchanger. They found that an increase in the gap velocity could reduce the fouling rate on the surface of plate heat exchanger.

In the present section, deposit formation on the surface of plate heat exchangers in different Egyptian fertilizer plants will be investigated. The effect of heat exchanger geometry and flow patterns on the fouling behavior will be shown.

2.2 Process description

Ammonia is the basic raw material in urea production. Ammonia plants in question operate using Uhde's proprietary ammonia process that is based on the well-established Haber-Bosch process. In the first stage, the raw material natural gas is desulphurized, then cracked into its individual chemical components catalytically by adding steam to generate the hydrogen required for ammonia synthesis. This process also generates carbon monoxide, carbon dioxide, hydrogen and residues of methane from the natural gas cracking process. In the next stage nitrogen is added to the process by combusting methane, CO and H_2 using air. With the addition of steam, carbon monoxide is converted to CO_2 using catalytic converters and then scrubbed out of the synthesis gas formed. The selectively scrubbed CO_2 is fed into the urea processing plant as the process medium together with the produced ammonia as starting material. The urea plants operate using the Stamicarbon process that was developed in the Netherlands [Uhde].

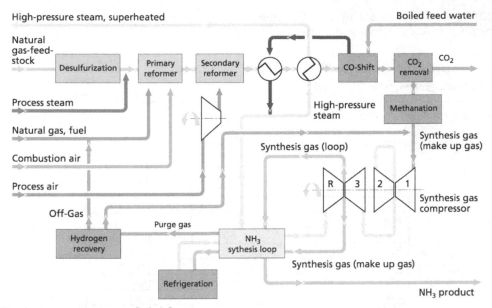

Fig. 3. Ammonia process [Uhde].

In the CO_2 scrubbing process three plate heat exchangers are switched in parallel, two in operation (A and B) and one in standby (C). Figure 4 shows the three coolers with their operating conditions. The CO_2 flows into the PHEs as a gas-steam mixture at 94 °C and is cooled down in a countercurrent process to 33 °C. Water at 30 °C is used as coolant. Each of the 10 tons and 3 meter high PHEs has 1000 m² of high-performance stainless steel (1.4539; AISI 904 L) VT-plates. The transferred heat capacity is 14.5 megawatts.

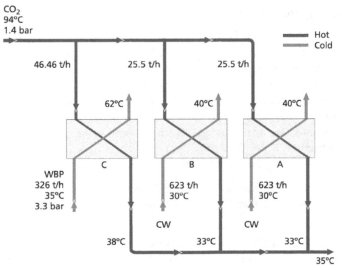

Fig. 4. CO_2 coolers used in the scrubbing process.

Nile river water treated by NALCO inhibitors is used in an open loop as the cooling medium for the CO_2 coolers, the specifications of the cooling water used is given in Table 1.

	CaH	Alkalinity	Chlorides	Inhibitors
Nile water	90 ppm	138	25 ppm	N-7356P: 30ppm, N-73203: 95ppm

Table 1. Cooling water specifications.

A typical analysis for Nile river water is shown in Table 2.

Substrate		Unit
Chloride	77.5	ppm
Ca	48	ppm
Mg	14.5	ppm
Na	60	ppm
K	9	ppm
Fe	0.1	ppm
SO$_4$	57.5	ppm
SiO$_2$	2	ppm
HCO$_3$	180	ppm
KMnO$_4$	10.1	ppm
Total hardness	172.5	ppm CaCO$_3$
TDS	380	ppm
pH	7.8	-
Alkalinity	180	ppm CaCO$_3$

Table 2. Nile river water analysis.

2.3 Problem description and observations

The cooling water flow rate on the CO_2 coolers (HP Scrubber) dropped from 500m³/hr to 300m³/hr due to fouling on the cooling water side, which caused operation problems in the Urea plant. The CO_2 outlet temperature was increasing with time and achieved about 50°C after 30 days of operation before the shutdown of the unit for mechanical cleaning. The CO_2 cooler was opened for mechanical cleaning; the PHE's inlet was plugged with plastic bags and pieces of bottles. Deposits were accumulated at an area about 20cm from the plate inlet and selectively covered the plate surface, as can be seen in Figure 5. They could plug the channels and restrict the water flow over the plate. These deposits accumulated due to the reduction of the gap velocity (shear stress) which increased the surface temperature.

A sample from the deposits was taken and analysed using ashing and X-ray Fluorescence (XRF). The sample was dried at 105 °C before ashing and XRF analysis. The results are shown in Table 3.

The ashing results showed that 14% of the sample was lost at a temperature below 500°C, which represents the organic material and can be considered as normal range. The XRF analysis showed that the main element in the deposits is zinc hydroxide as ZnO (38%) and

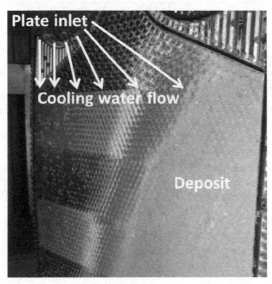

Fig. 5. Deposits formed on the surface of VT-plate.

the second is calcium phosphate (11%), which participated as a result of the increase of the plate surface temperature resulting from the reduction in the cooling water flow rate.

Loss at 500°C	14 %
Loss at 925°C	23 %

(a)

Substrate	Mass %
Magnesium (MgO)	3
Aluminium (Al_2O_3)	1
Silicon (SiO_2)	2
Phosphorous (P_2O_5)	20
Sulphur (SO_3)	1
Calcium (CaO)	11
Iron (Fe_2O_3)	1
Zinc (ZnO)	38
Total oxides (normalized to loss 925 °C)	77

(b)

Table 3. (a) Ashing results, (b) Elemental analysis as oxides using XRF.

2.4 Technical solutions

2.4.1 Redesigning of the PHEs

The surface area of the CO_2 cooler was reduced by removing 86 plates out of 254 plates (the surface area was reduced by 34%). The average cooling water velocity inside the gaps was increased from 0.30 to 0.42 m/s, as can be seen in Table 4.

	Original Design	After modification
Plates number	254	168
Gap velocity [m/s]	0.30	0.42
Surface tension [Pa]	5.31	10.84
Reynolds number	3259	4599
Surface temperature [°C]	72	69

Table 4. Design modification for CO_2 cooler in Helwan fertilizer plant, Egypt.

The deposits formed on the surface of the plates were decreased as a result of the increase in the shear stress and the decrease of the surface temperature from 72 to 69°C. The surface temperature was calculated from the fluids temperatures, thermal conductivities and duties on both sides. The operation time for the cooler was increased from 30 days to 43 days and the plates were cleaned after more than 40 days of operation, as shown in Figure 6.

The CO_2 outlet temperature started to increase after about 23 days of operation due to the accumulation of deposits on the cooling water side which led to a reduction in the cooling water flow rate. The unit was opened after about 43 days for mechanical cleaning.

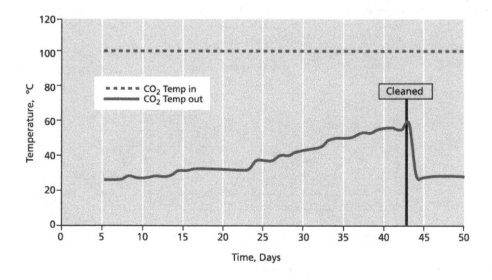

Fig. 6. Inlet and outlet CO_2 temperatures as a function of time.

2.4.2 New plate geometry

A new cooler with computer-modeled plate geometry from the NT (New Technology) series was installed in parallel with the existing two coolers. The NT Series sets new economic standards with low investment costs, operation and maintenance. The optimized OptiWave plate design requires less heat transfer surface for the same performance. The new EcoLoc gaskets and installation methods simplify maintenance and ensure a perfect fit of the gasket and plate packs. The new plates have the advantage of higher gap velocities (shear stress) due to better fluid distribution over the plates and smaller gap size.

The advantages of the NT-plates at a glance:

- High heat transfer rates
- Low investment and service costs
- Optimized distribution of media
- Simplified handling
- Quick and safe gasket replacement
- Flexible solutions for special requirements
- Non-standard materials available
- Leading manufacturer's know-how

In conventional plates the fluid velocity over the plate's width is decreasing, the more the fluid is distributed from the inlet over the whole plate width. This is due to the higher pressure drop in longer flow channels. The optimized fluid distribution channels of the NT series lead to balanced velocity over the whole plate width and an equal distribution of the medium (Figure 7).

The flow channels of the NT-plates vary in their width and were optimized based on Computational Fluid Dynamics (CFD). The channels located further away from the inlet hole have bigger diameter than those closer to the inlet hole.

Fewer deposits were accumulated on the NT-plates due to the asymmetric flow distribution over the channels as can be seen in Figure 8. These deposits were formed because the unit was taken into operation in parallel with the old two VT-plates units and hence most of the cooling water was flowing inside them. The NT-plates unit was designed in principle to replace one of the VT-plates units so that the gap velocity could be increased.

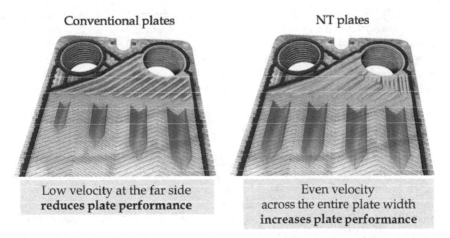

Conventional plates

Low velocity at the far side
reduces plate performance

NT plates

Even velocity
across the entire plate width
increases plate performance

Fig. 7. Velocity distribution over the NT-plate compared with conventional plates.

Fig. 8. Deposits formed on the surface of NT-plate.

2.5 Conclusions

Nile water treated with NALCO inhibitors caused fouling problems inside CO_2 coolers in different Egyptian ammonia plants. Technical solutions including redesigning of the PHEs and new plate geometries were investigated. Reducing the surface area of the CO_2 coolers by 34% increased the gap velocity from 0.30 to 0.42 m/s (shear stress from 5.31 to 10.84 Pa) and hence decreased fouling. The operation time for the cooler was increased from 30 days to 43 days. NT-plates with asymmetric flow distribution over the channels decreased the rate of deposition on the surface of the plates.

3. Solving fouling problems by surface modification

In a recent study, Nano-composite coatings were used to reduce fouling inside gasketed plate heat exchangers involved in food production. An antifouling coating with low surface energy (low wettability) led to a hydrophobic and oleophobic effect. The goal of the project was the application of new surface coatings (nanotechnology) to avoid or minimize adhesion, improve process management, simplify cleaning processes with lesser resources and chemical use, and increase the product reliability.

The test facility constructed by the Institute of Environmental Process Engineering (IUV) at the University of Bremen in Germany used for the investigation of milk adhesion and the stability of the coatings on small cylindrical ducts. A number of coatings and surface treatments were tested. A pilot plant including a milk pasteurizer at the Institute of Food Quality LUFA Nord-West in Oldenburg-Germany was used for the thermal treatment of whey protein solutions. Heat exchanger plates coated with different nano-composites as well as electropolished plates installed in the heating section of the pasteurizer were tested. Significant differences were observed between coated and uncoated plates. The coated plates showed reduced deposit buildup in comparison with the uncoated stainless steel plates. Polyurethane-coated plates exhibited the thinnest deposit layer. Electro-polished plates also reduced deposit buildup in comparison to the standard stainless steel plates and were almost comparable to the coated plates. The time required for cleaning in place (CIP) with the coated plates was reduced by 70% compared to standard stainless steel plates.

3.1 Introduction

Production problems, like decrease of production rate and increase in the intensity of cleaning procedure, arise in the dairy industry as a result of the deposit adhesion to the plate surface. The deposits must be removed by regular and intensive cleaning procedures in order to comply with hygiene and quality regulations for the dairy industry (Augustin et al., 2007). If not controlled carefully, deposits can cause deterioration in the product quality because milk cannot be heated up to the required pasteurization temperature. Milk deposits generally form so fast that heat exchangers must be cleaned regularly to maintain production efficiency and meet strict hygiene standards and regulations (Bansal and Chen, 2006). Energy losses, lost productivity, manpower and cleaning expenses cause immense costs (Beuf et al., 2003). In the dairy industry, fouling and the resulting cleaning of the process equipment account for about 80% of the total production costs (Bansal and Chen, 2006).

Gasketed plate heat exchangers with stainless steel plates are commonly used in the dairy industry. Stainless steel surfaces have high surface energies. The adhesion of product on solid surfaces is determined by the surface roughness and surface energy. The adhesion of deposits could be reduced by either decreasing the surface energy of the metal or by coating the metal surface with high anti-adhesion effect (low surface energy) materials, such as those made of nanoparticles (Gerwann et al., 2002). The application of nano-coatings with their anti-adhesion effects reduces the buildup of deposits on the surface of heat exchanger plates due to the reduction of adhesive forces. The operation efficiency of the plant can be significantly improved and the general hygienic situation of the product can increase. Additionally, intensity and frequency of cleaning can be substantially reduced to achieve the desired degree of product quality (Kück et al., 2007).

Beuf et al. (2003) studied the fouling of dairy product on modified stainless steel surfaces in a plate and frame heat exchanger. Different surface modifications, such as coatings (diamond like carbon [DLC], silica, SiOX, Ni-P-PTFE, Excalibur, Xylan) and ion implantation (SiF+, MoS2) were analyzed. No significant difference was found between the modified stainless steels and the unmodified one. The cleaning efficiency of plates coated with Ni-P-PTFE was the best. The experimental results of Zhao et al. (2007) showed that the surface free energy of the Ni–P–PTFE coating had a significant influence on the adhesion of bacterial, protein and mineral deposits. The Ni–P–PTFE coating reduced the adhesion of these deposits significantly.

The fouling behavior of whey protein solutions on modified stainless steel (SS) surfaces coated with diamond-like carbon (DLC) and titanium nitride (TiN) have been studied by Premathilaka et al. (2007). They concluded that fouling decreased in the order DLC > SS > TiN and cleaning time decreased in the order TiN > SS > DLC.

The goal of the present work is to assess new surface coatings (developed by the Institute of New Materials, INM, in Germany) with low surface energy and low roughness to avoid or minimize adhesion of deposits, simplify cleaning processes, reduce resource and chemical requirements, and increase product quality and consistency. The work will assess the deposit buildup during the thermal treatment of milk.

3.2 Experimental

3.2.1 Coated surfaces

The anti-adhesion nano-composite coatings, with hydrophobic and oleophobic effectiveness, used in this work were produced from commercially available polymer matrices such as epoxy, polyurethane or Polyamide systems which were reactively cross-linked with per-fluorinated monomers (or oligomers) and ceramic reinforcement particles. The coating material application was similar to wet chemical coating by spraying, and the required mechanical properties were obtained through a thermal cross linking step. The epoxy and polyurethane systems were hardened at 130 ° C for 1 hour, while the Polyamide systems were hardened at 200 °C for 2 hours, in order to ensure an optimum layer formation. Table 5 summarizes the plates used and their specifications.

Plate	Material	Contact angle Water [°]	Contact angle Milk [°]	Surface roughness [μm]
U1	Stainless steel	83.8	69.6	0.80
U1min,e	Electrically-polished stainless steel for one minute	61.7	93.5	0.50
U5min,e	Electrically-polished stainless steel for five minutes	60.3	87.8	0.22
A1	Epoxy-resin based coating of INM	97.6	94.5	0.92
A2	Epoxy-resin based coating of INM	91.0	97.6	0.95
A9	Polyurethane based coating of INM	92.2	88.3	0.23
A10	Polyurethane based coating of INM	93.4	95.5	0.06
A17	Epoxy-resin based coating of INM	95.8	95.1	1.14

Table 5. Contact angle and surface roughness of the sheets prepared by INM.

3.2.2 Laboratory scale testing

A laboratory facility was constructed by the Institute of Environmental Process Engineering (IUV), University of Bremen in Germany, for the investigation of milk adhesion. A heat exchanger was designed to enable thermal and hydraulic load measurements with variable designs. Its principal components were a double wall heated receiver tank, controllable pump, electromagnetic flow meter and the test cell (duct). A closed loop recirculation configuration was used to decrease the volume of the test medium required. Furthermore, a fast sample change by simple removal of the test cell (duct) was performed. Figure 9 shows the laboratory apparatus used and the test duct. The test channel employs an annular geometry, where the inner cylinder is engaged with an electric heater. The middle part (with the threaded ends) is made of stainless steel, while the coin section (right) and the fastener (left) are made from a high-performance plastic Polychlorotriflouroethylene (PCTFE). This arrangement allows the middle section, which incorporates the heater, to be heated by thermal conduction without large heat losses. The coating material is applied to a small stainless steel tube which is pushed over the heater.

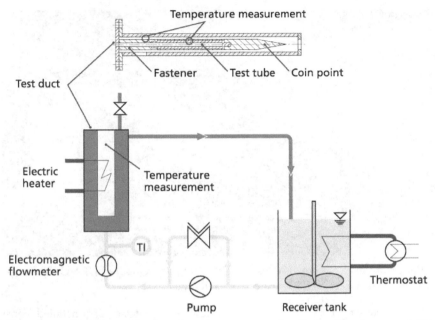

Fig. 9. Flowchart of laboratory heat exchanger apparatus (Institute of Environmental Process Engineering IUV, Universität Bremen).

For the experiments, a 10% (by weight) aqueous whey protein solution was set in the receiver tank. The solution was prepared by solving a whey protein concentrate WPC35 in water until the required concentration was obtained. The pH was adjusted to 6.0 using a 0.1 mol/liter HCl solution. Pre-heating was carried out to about 43 °C. The solution was pumped in the closed cycle of the experimental setup, the electric heater of the test channel was activated and the measuring procedure was started. After each trial, the whey protein solution was replaced to exclude any effect of heating on the ingredients. After each run, the tube was cleaned with 0.1 molar NaOH solution with cross flow velocity of 0.6 m/s. The experimental parameters were:

Volumetric flow rate: 0.036 – 0.37 m3/h
Whey protein concentration: 10% (by weight)
Average flow velocity in annulus: 0.2 m/s
Fluid temperature (measuring section): 45 °C
Temperature of the heating element: 230 °C
Heat flux: 20 kW/m²
Experimental time: 15 to 30 min.

3.2.3 Pilot plant testing

Industrial tests with milk were carried out on a small plant by the Institute of Food Quality LUFA-Oldenburg-Germany, with the support of the company GEA PHE Systems (Figure 10). The pilot plant can produce almost all dairy products. It is used for training purposes as well as technological support and procedure development to the food industry.

Fig. 10. Pilot plant used for practical tests (Institute of Food Quality LUFA-Oldenburg-Germany) with GEA Ecoflex VT04 plate heat exchanger.

The plate heat exchanger, in which coated and uncoated plates can be installed, consists of two cooling sections (deep cooler with 8 plates and pre-cooler with 10 plates), heat recovery section (with 12 plates), heating section (with 7 plates) and hot water section (with 6 plates). Before assembling the heat exchanger, selected plates in the heating and heat recovery sections were coated using the method described in section 3.2.1. As a reference, stainless steel, electro-polished and PTFE coated plates were also installed in the heat exchanger (Figure 11). Table 6 details the samples used and their specifications.

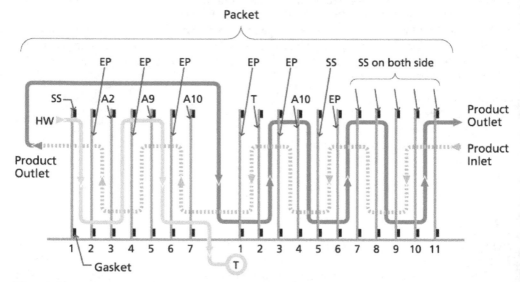

Fig. 11. Plates layout inside GEA Ecoflex VT04 plate heat exchanger.

Substrate	Material	Thickness [μm]
SS	Stainless steel	-
EP	Electrically-polished stainless steel	-
A2	Epoxy-resin based coating of INM	83.7
A9	Polyurethane based coating of INM	53.0
A10	Polyurethane based coating of INM	85.2
A67	Polyurethane based coating of INM	27.6
PTFE	Teflon	22.5

Table 6. Samples specifications used in the pilot plant experiments by LUFA.

The milk was pumped from the receiver tank through the pasteurizer at a constant flow rate. The process steps of heating, cooling and heat recovery were combined together. After a working time of 4 hours, the test was stopped and the plates of the heater and heat recovery sections were removed in order to measure deposit formation. Visual observations and mass investigations were done. Furthermore, the cleaning effectiveness was assessed.

3.3 Results and discussion

Technical investigations were carried out by IUV and LUFA on the deposits formed from whey protein solution in both the laboratory facility and the pilot plant.

3.3.1 Laboratory tests by IUV

Laboratory investigations were carried out by IUV on the deposit of whey protein on the tube surface. Different stainless steel tubes were tested by IUV using the laboratory heat exchanger apparatus described in section 3.2. Figure 12 shows the deposit accumulation rates of whey protein solution for the different tube surfaces.

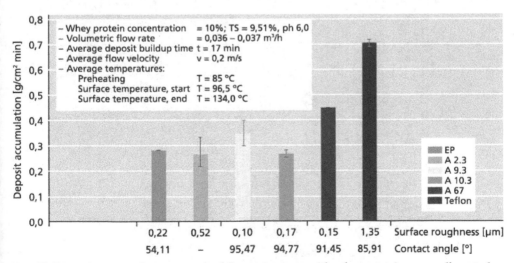

Fig. 12. Deposit accumulation rates for laboratory tests with whey protein on small coated cylindrical ducts. Plate characteristics are given in Table 6.

The Polyurethane-coated tubes gave the thinnest deposit layer, closely followed by the electropolished tubes. The laboratory cleaning tests showed that under the same hydrodynamic conditions, the cleaning time for test tube A9 is only 20% of that needed for the standard stainless steel tube.

3.3.2 Pilot plant tests by LUFA

In a test series LUFA Nord-West in Oldenburg-Germany examined the formation of deposit on test PHE plates which had undergone different treatments. Different coated plates were installed in the heating section of a pasteurizer, with PTFE coated plates next to electro-polished and standard stainless steel plates. The anti-fouling coatings were high-molecular polymers with implanted nano-particles which resulted in high hardness and scratch resistance. The pasteurizer was operated with a 10% (by weight) whey protein solution which was heated up to 85°C. Figure 13 shows the amount of residue, in g, for different surfaces in three tests. It is noteworthy that in these test conditions there is significant whey protein deposition on uncoated, electro-polished and A2-coated stainless steel.

The coatings A2 and A10 showed reduced deposit buildup (the PTFE coating gave more deposit buildup than the standard stainless steel plate). The plates coated with A10 coating had the lowest adhesion, which was similar to the laboratory test results. The deposit buildup on the electro-polished plates was lower than the standard stainless steel plate and almost comparable to the coated plates. Cleaning studies indicated that the cleaning in place (CIP) time, for all coatings was shorter than that for the standard stainless steel plate: PTFE coated plates down by 90%; coated plates down by 70%; electro polished plates down by 36%.

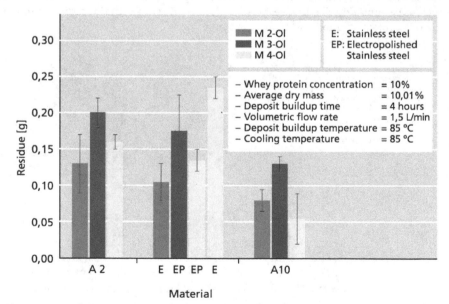

Fig. 13. Amount of deposits formed using whey protein solution, in three tests (m2, m3 and m4).

Figure 14 shows photographs for two different coated plates from the heating section in the heat exchanger after different experimental runs (A2 on left and A10 on right).

It is evident that the coatings have been locally destroyed at the contact points, as pointed by the red circles in the figure. The flow in the plate gap causes relatively high vibrations with particularly strong stresses to the contact points, which is added to the high thermal stresses. The coatings at the present stage of development could not withstand these stresses and need further development.

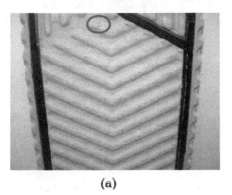 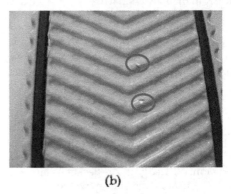

(a)   (b)

Fig. 14. Coated heat exchanger plates from heating section: (a) A2 coating, (b) A10 coating.

3.4 Conclusions

Nano-composites could be used as anti-fouling coatings to decrease fouling inside gasketed plate heat exchangers for the dairy industry. Industrial tests showed that the coatings A2 and A10 reduced fouling, though the PTFE coating showed higher fouling than the standard stainless steel plate. The deposit buildup on the electro-polished plates was lower than the standard stainless steel plates and almost comparable to the coated plates. A CIP time reduction was observed for all coatings: PTFE coated plates down by 90%; nano-composites coated plates down by 70%; electro polished plates down by 36%. Pilot plant testing indicated the coatings must be further developed so that they can withstand the thermal and mechanical stresses which arise in industrial operation.

4. Nomenclature

CIP	Cleaning in place
EP	Electrically-polished stainless steel
INM	Institute of New Materials
IUV	Institute of Environmental Process Engineering
LUFA	Institute of Food Quality
NT	New technology
PCTFE	Polychlorotrifluorethylene
PHE	Plate heat exchanger
PTFE	Polytetrafluorethene (Teflon)
SS	Stainless steel
VT	Varitherm
XRF	X-ray Fluorescence

5. References

Augustin, W., Geddert, T., Scholl, S. (2007). Surface treatment for the mitigation of whey protein fouling, *Proceedings of 7th International Conference on Heat Exchanger Fouling and Cleaning*, pp. 206-214, ECI Symposium Series, Volume RP5Tomar, Portugal.

Bani Kananeh, A., Scharnbeck, E., Kück, U. D. (2009). Application of antifouling surfaces in plate heat exchanger for food production, *Proceedings of 8th International Conference on Heat Exchanger Fouling and Cleaning*, pp. 154-157, Schladming, Austria.

Bani Kananeh, A., Scharnbeck, E., Kück, U.D. and Räbiger, N. (2010). Reduction of Milk Fouling Inside Gasketed Plate Heat Exchanger Using Nano-Coatings. *Food and Bioproducts Processing*, Vol. 88, No. 4, (December 2010), pp. 349-356.

Bansal, B., Chen, X. D. (2006). A critical review of milk fouling in heat exchangers, *Comprehensive Reviews in Food Science and Food Safety*, Vol. 5, No. 2, (April 2006), pp. 27-33.

Beuf, M., Rizzo, G., Leuliet, J.C., Müller-Steinhagen, H., Yiantsios, S., Karabelas, A., Benezech, T. (2003). Fouling and cleaning of modified stainless steel plate heat exchangers processing milk products, *Proceedings of ECI Conference on Heat Exchanger Fouling and Cleaning: Fundamentals and Applications*, pp. 99-106, Vol. RP1, Article 14, Santa Fe, New Mexico, USA.

Delplace, F., Leuliet, J. C. and Tissier, J.P. (1994). Fouling experiments of a plate heat exchanger by whey proteins solutions, *Transactions on IChemE C 72*, pp.163–9.

Gerwann, J., Csögör, Z., Becker-Willinger, C. and Schmidt, H. (2002). Antimicrobic low surface-free energy nanocomposite coatings for medical applications, *Proceedings of Hygienic Coatings Conference*, Brussels.

Kukulka, D. J. and Leising, P. (2009). Evaluation of Surface Coatings on Heat Exchangers. *Chemical Engineering Transactions*, Vol.18, (May 2009), pp. 339-344.

Kück, U. D., Hartmann, D., Manske, S., Kück, A. and Räbiger, N. (2007). Entwicklung neuer Verarbeitungsprozesse für die Lebensmittelherstellung durch Anwendung von neuartigen funktionalen Materialoberflächen, *AiF-Abschlußbericht (AiF-Nr. 14228N/1)*, Bremen.

Müller-Steinhagen, H. (1999). Cooling-Water Fouling in Heat Exchangers. *Advances in Heat Transfer*, Vol. 33, (1999), pp. 415-496.

Müller-Steinhagen, H. (2006). Verschmutzung von Wärmeübertragerflächen. In: *VDI-Wärmeatlas Ausgabe 2006*, Verein Deutscher Ingenieure and VDI- Gesellschaft Verfahrenstechnik und Chemieingenieurwesen (GVC), pp. (Od1-Od30), Springer-Verlag Berlin Heidelberg, Germany.

Pana-Suppamassadu, K., Jeimrittiwong, P., Narataruksa, P. and Tungkamani, S. (2009). Effects of Operating Conditions on Calcium Carbonate Fouling in a Plate Heat Exchanger. *World Academy of Science, Engineering and Technology*, No. 53, (May 2009), pp. 1204-1215.

Premathilaka, S. S., Hyland, M. M., X.D. Chen, X. D., Watkins, L. R., Bansal, B. (2007). Interaction of whey protein with modified stainless steel surfaces, *Proceedings of 7th International Conference on Heat Exchanger Fouling and Cleaning*, pp. 150-121 , ECI Symposium Series, Volume RP5, Article 21,Tomar, Portugal.

Uhde GmbH, Available from: <http://www.uhde.eu/index_flash.en.epl>

Wang, L., Sunden, B, Manglik, R.M. (2007). *Plate Heat Exchangers: Design, Applications and Performance*, WIT Press, ISBN 978-1-85312-737-3, Southampton, Great Britain.

Zhao, Q., Liu, C., Liu, Y., Wang, S. (2007). Bacterial and protein adhesion on Ni-P-PTFE coated surfaces, *Proceedings of 7th International Conference on Heat Exchanger Fouling and Cleaning*, pp. 237-242, ECI Symposium Series, Volume RP5, Article 33, Tomar, Portugal.

Permissions

The contributors of this book come from diverse backgrounds, making this book a truly international effort. This book will bring forth new frontiers with its revolutionizing research information and detailed analysis of the nascent developments around the world.

We would like to thank Prof. Dr. Ing. Jovan Mitrovic, for lending his expertise to make the book truly unique. He has played a crucial role in the development of this book. Without his invaluable contribution this book wouldn't have been possible. He has made vital efforts to compile up to date information on the varied aspects of this subject to make this book a valuable addition to the collection of many professionals and students.

This book was conceptualized with the vision of imparting up-to-date information and advanced data in this field. To ensure the same, a matchless editorial board was set up. Every individual on the board went through rigorous rounds of assessment to prove their worth. After which they invested a large part of their time researching and compiling the most relevant data for our readers. Conferences and sessions were held from time to time between the editorial board and the contributing authors to present the data in the most comprehensible form. The editorial team has worked tirelessly to provide valuable and valid information to help people across the globe.

Every chapter published in this book has been scrutinized by our experts. Their significance has been extensively debated. The topics covered herein carry significant findings which will fuel the growth of the discipline. They may even be implemented as practical applications or may be referred to as a beginning point for another development. Chapters in this book were first published by InTech; hereby published with permission under the Creative Commons Attribution License or equivalent.

The editorial board has been involved in producing this book since its inception. They have spent rigorous hours researching and exploring the diverse topics which have resulted in the successful publishing of this book. They have passed on their knowledge of decades through this book. To expedite this challenging task, the publisher supported the team at every step. A small team of assistant editors was also appointed to further simplify the editing procedure and attain best results for the readers.

Our editorial team has been hand-picked from every corner of the world. Their multi-ethnicity adds dynamic inputs to the discussions which result in innovative outcomes. These outcomes are then further discussed with the researchers and contributors who give their valuable feedback and opinion regarding the same. The feedback is then collaborated with the researches and they are edited in a comprehensive manner to aid the understanding of the subject.

Apart from the editorial board, the designing team has also invested a significant amount of their time in understanding the subject and creating the most relevant covers. They scrutinized every image to scout for the most suitable representation of the subject and create an appropriate cover for the book.

The publishing team has been involved in this book since its early stages. They were actively engaged in every process, be it collecting the data, connecting with the contributors or procuring relevant information. The team has been an ardent support to the editorial, designing and production team. Their endless efforts to recruit the best for this project, has resulted in the accomplishment of this book. They are a veteran in the field of academics and their pool of knowledge is as vast as their experience in printing. Their expertise and guidance has proved useful at every step. Their uncompromising quality standards have made this book an exceptional effort. Their encouragement from time to time has been an inspiration for everyone.

The publisher and the editorial board hope that this book will prove to be a valuable piece of knowledge for researchers, students, practitioners and scholars across the globe.

List of Contributors

Muthuraman Subbiah
Higher College of Technology, Oman

Mohamad Kharseh
Willy´s CleanTech AB, PARK 124 Karlstad, Sweden

Pablo Dolado, Ana Lázaro, José María Marín and Belén Zalba
University of Zaragoza / I3A – GITSE, Spain

Béatrice A. Ledésert and Ronan L. Hébert
Géosciences et Environnement Cergy, Université de Cergy-Pontoise, France

Dick G. Klaren and Eric F. Boer de
KLAREN BV, The Netherlands

S. N. Kazi
Department of Mechanical and Materials Engineering, Faculty of Engineering, University of Malaya, Kuala Lumpur, Malaysia

Ali Bani Kananeh and Julian Peschel
GEA PHE Systems, Germany

Printed in the USA
CPSIA information can be obtained
at www.ICGtesting.com
JSHW011400221024
72173JS00003B/355